THE CHILDHOOD ENVIRONMENT AND ADULT DISEASE

The Ciba Foundation is an international scientific and educational charity. It was established in 1947 by the Swiss chemical and pharmaceutical company of CIBA Limited—now CIBA-GEIGY Limited. The Foundation operates independently in London under English trust law.

The Ciba Foundation exists to promote international cooperation in biological, medical and chemical research. It organizes about eight international multidisciplinary symposia each year on topics that seem ready for discussion by a small group of research workers. The papers and discussions are published in the Ciba Foundation symposium series. The Foundation also holds many shorter meetings (not published), organized by the Foundation itself or by outside scientific organizations. The staff always welcome suggestions for future meetings.

The Foundation's house at 41 Portland Place, London W1N 4BN, provides facilities for meetings of all kinds. Its Media Resource Service supplies information to journalists on all scientific and technological topics. The library, open five days a week to any graduate in science or medicine, also provides information on scientific meetings throughout the world and answers general enquiries on biomedical and chemical subjects. Scientists from any part of the world may stay in the house during working visits to London.

Ciba Foundation Symposium 156

THE CHILDHOOD ENVIRONMENT AND ADULT DISEASE

A Wiley-Interscience Publication

1991

JOHN WILEY & SONS

Chichester · New York · Brisbane · Toronto · Singapore

Published in 1991 by John Wiley & Sons Ltd.
Baffins Lane, Chichester
West Sussex PO19 1UD, England

Other Wiley Editorial Offices

John Wiley & Sons, Inc., 605 Third Avenue,
New York, NY 10158-0012, USA

Jacaranda Wiley Ltd, G.P.O. Box 859, Brisbane,
Queensland 4001, Australia

John Wiley & Sons (Canada) Ltd, 22 Worcester Road,
Rexdale, Ontario M9W 1L1, Canada

John Wiley & Sons (SEA) Pte Ltd, 37 Jalan Pemimpin 05-04,
Block B, Union Industrial Building, Singapore 2057

Suggested series entry for library catalogues:
Ciba Foundation Symposia

Ciba Foundation Symposium 156
viii + 243 pages, 37 figures, 20 tables

Library of Congress Cataloging-in-Publication Data
The Childhood environment and adult disease.
    p. cm.—(Ciba Foundation symposium; 156)
    Based on the Symposium on the Childhood Environment and Adult
Disease, held at the Ciba Foundation, London, 15–17 May 1990.
    Editors: Gregory R. Bock (organizer) and Julie Whelan.
    'A Wiley–Interscience publication.'
    Includes bibliographical references and index.
    ISBN 0 471 92957 3
    1. Environmentally induced diseases in chldren—Complications and
sequelae—Congresses.   2. Diseases—Causes and theories of
causation—Congresses.   3. Mental illness—Etiology—Congresses.
4. Child development—Congresses.   I. Bock, Gregory.   II. Whelan,
Julie.   III. Symposium on the Childhood Environment and Adult
Disease (1990: Ciba Foundation)   IV. Series.
    [DNLM: 1. Child Development—congresses.   2. Disease—etiology—
congresses.   3. Environmental Exposure—congresses.   4. Fetal
Development—congresses.   5. Human Development—congresses.
6. Social Environment—congresses.   W3 C161F v. 156/WS 103 C536
1990]
RJ383.C49   1991
618.92'98 – dc20
DNLM/DLC
for Library of Congress   90-13144
                                                          CIP

British Library Cataloguing in Publication Data
The childhood environment and adult disease.
    1. Man. Diseases. Causes
    I. Bock, Gregory R.   II. Whelan, Julie.   III. Series
    616.071

    ISBN 0 471 92957 3

Phototypeset by Dobbie Typesetting Limited, Tavistock, Devon.
Printed and bound in Great Britain by Biddles Ltd., Guildford.

Contents

Participants

D. J. P. Barker (*Chairman*)   MRC Environmental Epidemiology Unit, University of Southampton, Southampton General Hospital, Southampton SO9 4XY, UK

C. Blakemore   University Laboratory of Physiology, University of Oxford, Parks Road, Oxford OX1 3PT, UK

P. Casaer   Division of Paediatric Neurology & Developmental Neurology Research Unit, Department of Paediatrics, University Hospital Gasthuisberg, B-3000-Leuven (Louvain), Belgium

A. Caspi   Department of Psychology, University of Wisconsin-Madison, W J Brogden Psychology Building, 1201 West Johnson Street, Madison, WI 53706, USA

R. K. Chandra   Department of Paediatrics, Medicine & Biochemistry, Memorial University of Newfoundland, and Janeway Child Health Centre, Janeway Place, St John's, Newfoundland, Canada A1A 1R8

J. Dobbing   Department of Child Health, University of Manchester, Medical School, Stopford Building, Oxford Road, Manchester M13 9PT, UK

J. Golding   Department of Child Health, Royal Hospital for Sick Children, St Michael's Hill, Bristol BS2 8BH, UK

M. Hamosh   Division of Developmental Biology & Nutrition, Department of Paediatrics, Georgetown University Children's Medical Center, 3800 Reservoir Road NW, Washington DC 20007-2197, USA

M. Hanson   Fetal & Neonatal Research Group, Department of Obstetrics & Gynaecology, University College London, 86–96 Chenies Mews, London WC1E 6HX, UK

J. K. Lloyd   Department of Child Health, Institute of Child Health, 30 Guilford Street, London WC1N 1EH, UK

A. Lucas MRC Dunn Nutrition Unit, Downhams Lane, Milton Road, Cambridge CB4 1XJ, UK

C. N. Martyn MRC Environmental Epidemiology Unit, University of Southampton, Southampton General Hospital, Southampton SO9 4XY, UK

T. W. Meade MRC Epidemiology & Medical Care Unit, Northwick Park Hospital, Watford Road, Harrow, Middlesex HA1 3UJ, UK

G. E. Mott Department of Pathology, University of Texas Health Science Center, 7703 Floyd Curl Drive, San Antonio, TX 78284-7750, USA

E. R. Moxon Department of Paediatrics, John Radcliffe Hospital, Headington, Oxford OX3 9DU, UK

R. M. Murray Institute of Psychiatry & King's College Hospital, De Crespigny Park, Denmark Hill, London SE5 8AF, UK

J. Parnas Department of Psychiatry, Institute of Psychology, University of Copenhagen, Kommunehospital, DK-1399 Copenhagen K, Denmark

M. P. M. Richards Child Care & Development Group, University of Cambridge, Free School Lane, Cambridge CB2 3RF, UK

M. L. Rutter MRC Child Psychiatry Unit, Department of Child & Adolescent Psychiatry, Institute of Psychiatry, De Crespigny Park, Denmark Hill, London SE5 8AF, UK

J. L. Smart Department of Child Health, University of Manchester, Medical School, Stopford Building, Oxford Road, Manchester M13 9PT, UK

S. J. Suomi Laboratory of Comparative Ethology, National Institute of Child Health & Human Development, Building 31, Room B2B15, National Institutes of Health, 9000 Rockville Pike, Bethesda, MD 20892, USA

K. L. Thornburg Department of Physiology, The Oregon Health Sciences University, 3181 SW Sam Jackson Park Road, Portland, OR 97201, USA

M. E. J. Wadsworth MRC National Survey of Health & Development, University College & Middlesex Hospital Medical School, Department of Community Medicine, 66/72 Gower Street, London WC1E 6EA, UK

C. B. S. Wood Joint Academic Department of Child Health, The Medical Colleges of St Bartholomew's & The London Hospitals, Queen Elizabeth Hospital for Children, Hackney Road, London E2 8PS, UK

Introduction

D. J. P. Barker

MRC Environmental Epidemiology Unit, University of Southampton, Southampton General Hospital, Southampton SO9 4XY, UK

We are assembled to talk about the influence of the childhood environment on adult diseases, and we shall be covering four main areas. The first concerns mechanisms operating in early life which could have a bearing on cardiovascular disease. The second encompasses the interaction of nutrition with the immune system and the long-term effects of infection in childhood. In the third, we shall consider brain growth at critical periods of development and some exciting new ideas about schizophrenia, suggesting that it arises as a consequence of damage to the brain around the time of birth. Finally, we shall move into the area of psychosocial development.

As a group, we are remarkably heterogeneous, necessarily so, and none of us can know much about what will be discussed outside our particular fields. I am happy to admit that my knowledge of pre-alpha cell clustering is quite limited, and when it comes to the species-normative maternal rearing of rhesus monkeys, I am innocent! We must therefore, throughout the symposium, make sure that we all understand the language being used in each area, for the benefit both of our discussion, and of the readers of the book which will be produced.

Where do we expect to get to in this symposium? We cannot, of course, know; but my hope is that we shall become aware that there are a number of areas where the importance of what happens in childhood is much greater than we have previously supposed. We are going to hear, for example, that diet in early life may affect one's lifetime expectation of allergic disease, that schizophrenia and motor neuron disease may originate in infancy, and speculation that the risk of dying from a stroke is essentially determined before birth. As we consider these exciting ideas, I hope we shall attain a sense of how much is known and how much is conjecture.

We will be thinking about mechanisms, and here there are some central concepts. The simplest is that if one really bad event happens in childhood, a major brain injury for example, it has immediate and irreversible consequences. From that we move to instances such as rheumatic heart disease where there is an event in childhood but only after a long interval are its harmful consequences apparent. In poliomyelitis, we have a model of diseases where the timing of the adverse event in childhood is critical in determining its consequences. It may be critical because an organ is at a critical stage of

1

development, or because the development of an entire function, such as personality or immunological competence, is at a critical stage.

When we discuss blood pressure we will meet the idea that the fetus, threatened by an adverse environment, may raise its blood pressure, which may be an effective response in terms of short-term survival, but may have as its price reduced long-term survival. One suspects there may be psychological analogies of this situation.

Another set of ideas relates to the consequences of infant feeding and social rearing practices. The message that is beginning to emerge is that infant feeding may set up metabolic patterns which determine responses to later challenges from the same stimulus—that is, high fat intake. Similarly, social rearing practices may determine responses to social challenges in adult life.

Professor Michael Rutter, when he talks about psychosocial development, will be introducing another set of mechanisms, a chain of adverse advents— not just the simplest chain, in which some people are especially unlucky and encounter one bad thing after another throughout life, but a more subtle chain in which people experiencing an adverse environment in early life become more likely to put themselves into an adverse environment later on. Even more intriguing is the idea that the interactions of personality with environment early on may lead somebody to *create* their own environment in adult life. There must be a wealth of points to discuss here.

The most easily awaited part of this three-day symposium will be my summing-up at the end. It seems unlikely that I will be able to condense our discussion of wide-ranging ideas into a few succinct sentences. I predict, however, that we shall agree that the environment in very early life is extremely important, and we shall add that this is an area that is seriously under-researched. The question is whether we shall have identified some obvious ways forward for research in particular areas, and whether there are concepts which unify research across the whole field.

We are now about to embark on a journey down many paths. Where will it take us?

The intrauterine environment and adult cardiovascular disease

D. J. P. Barker

MRC Environmental Epidemiology Unit, University of Southampton, Southampton General Hospital, Southampton SO9 4XY, UK

Abstract. Two recent findings suggest that maternal nutrition, and fetal and infant growth, have an important effect on the risk of cardiovascular disease in adult life. (1) Among 5225 men who were born in Hertfordshire, England during 1911–1930 and who were breast fed, those who had the lowest weights at birth and at one year had the highest death rates from cardiovascular disease. The differences were large and were reflected in differences in life expectancy. (2) In England and Wales there is a close geographical association between high death rates from cardiovascular disease, and poor maternal physique and health, and poor fetal growth. These findings raise the question of what processes link the intrauterine and early postnatal environment with risk of cardiovascular disease. Blood pressure, a known risk factor for cardiovascular disease, is one link. A recent study of 449 men and women now aged 50 showed that measurements at birth predicted blood pressure more strongly than current measures such as body mass. Levels of clotting factors in the blood and serum cholesterol (two other risk factors) may also be links.

1991 The childhood environment and adult disease. Wiley, Chichester (Ciba Foundation Symposium 156) p 3–16

There is increasing evidence that the intrauterine environment has an important effect on the risk of cardiovascular disease—that is, ischaemic heart disease and stroke—in adult life. This research originated in geographical studies. A puzzling aspect of the epidemiology of ischaemic heart disease and stroke in Britain is that they are more common in poorer areas and in lower income groups. The differences are large, greater than twofold (Gardner et al 1969, Registrar General 1978). For ischaemic heart disease they are also paradoxical, in that its steep rise in Britain and elsewhere has been associated with rising prosperity. Why should rates of ischaemic heart disease be lowest in the most prosperous places, such as London and the home counties? Variations in cigarette smoking and adult diet do not explain these differences.

We have examined the possibility that they result from geographical and social class differences in infant development 60 and more years ago. Past differences in infant development and health in England and Wales were reflected in the wide range of infant mortality. For example, in 1921–1925 infant mortality ranged from 44 per 1000 births in rural West Sussex to 114 in Burnley in Lancashire. The highest rates were generally in northern counties where large manufacturing towns had grown up around the coal seams. Rates were also high in poor rural areas such as north Wales. They were lowest in counties in the south and east, which have the best agricultural land and are historically the wealthiest (Local Government Board 1910).

We have used infant mortality statistics for England and Wales to compare the present distribution of adult death rates from cardiovascular disease with the past geographical distribution of different causes of infant mortality. These comparisons are made with the country divided into large towns and groupings of small towns and rural areas within counties, totalling 212 areas—a division of the country used in routine statistics since the turn of the century.

Figure 1 shows that the geographical pattern of death rates from cardiovascular disease closely resembles that of neonatal mortality (deaths before one month of age) in the past (Barker & Osmond 1986). At that time most neonatal deaths occurred during the first week after birth and were attributed

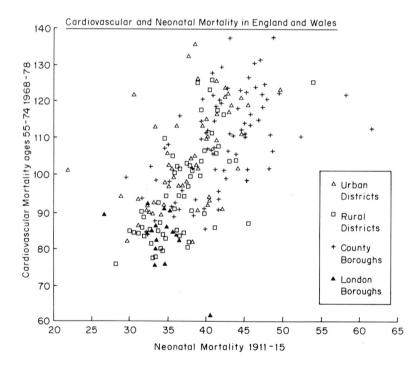

FIG. 1.   Standardized mortality ratios for cardiovascular disease (1968–1978) at ages 55–74, both sexes, and neonatal mortality 1911–1915, in the 212 areas of England and Wales.

to low birth weight (Local Government Board 1910). The geographical distribution of maternal mortality, from causes other than puerperal fever, was closely similar to neonatal mortality (Barker & Osmond 1987). Poor physique and health of the mothers was clearly implicated as a cause of high maternal mortality, and was partly a result of the poor nutrition and impaired growth of young girls (Campbell et al 1932). There is therefore a geographical association between high death rates from cardiovascular disease, poor fetal growth and poor maternal physique and health.

In addition to these associations, which indicate the importance of the intrauterine environment, the distribution of ischaemic heart disease is also related to post-neonatal mortality—deaths from one month to one year (Barker et al 1989a). Ischaemic heart disease, but not stroke, is therefore geographically linked to an adverse environment in infancy as well as in fetal life.

Mortality from stroke has fallen in Britain over the past 40 years (General Register Office 1911 et seq). This is consistent with past improvements in the intrauterine environment, as a result of improved maternal nutrition and physique. Ischaemic heart disease mortality, however, has risen steeply. It may therefore have two groups of causes, one acting through the mother and in infancy, and associated with poor living standards, the other acting in later life, and associated with affluence. This later influence seems likely to be linked to the high energy Western diet.

A recent follow-up study gave the first indication that the population associations of cardiovascular disease are also present in individuals. We have traced 5654 men born in Hertfordshire, England during 1911–1930 (Barker et al 1989b). From 1911 onwards, health visitors recorded the birth weights of all babies born in the county and visited their homes periodically throughout infancy. At one year the infant was weighed. The records of these visits have been preserved. Table 1 shows death rates from ischaemic heart disease according

TABLE 1 Standardized mortality ratios for ischaemic heart disease according to weight at one year in 5225 men who were breast fed

Weight at one year (lb)	Standardized mortality ratios	
≤18	112	(33)
19–20	81	(71)
21–22	100	(154)
23–24	69	(85)
25–26	61	(40)
≥27	38	(9)
All	81	(392)

Numbers of deaths in parentheses. 1 lb = 0.45 kg.
From Barker et al 1989b.

to weight at one year in the 5225 men who were breast fed. Hertfordshire is a prosperous part of England and rates of ischaemic disease are below the national average which, when rates are expressed as standardized mortality ratios (SMRs), is set as 100. Among men whose weights were 18 pounds or less at one year, death rates were around three times greater than in those who attained 27 pounds or more at one year. This is a strong relation: it spans more than 60 years, and it is graded. No similar relation was found in men who were bottle fed from birth, but the numbers were small. Similarly, the numbers of deaths from stroke in this initial sample are too few for analysis. The follow-up is being extended to 20 000 men and women.

Both prenatal and postnatal growth were important in determining weight at one year, since few infants with below average birth weights reached the heaviest weights at one. The lowest SMRs occurred in men who had above-average birth weight or weight at one year (Table 2). The highest SMRs were in men for whom birth weight was average or below and weight at one was below average. Among men for whom both weights were in the lowest group, 5.5 pounds or less and 18 pounds or less, the SMR was 220. The simultaneous effect of birth weight and weight at one year on death rates from ischaemic heart disease are shown in Fig. 2. The lines join points with equal risk of ischaemic heart disease. The values are risks relative to the value of 100 for those with average birth weight and weight at one.

From these findings we conclude that processes linked to growth and acting in prenatal or early postnatal life strongly influence risk of ischaemic heart disease. There is evidence that these processes include (1) the determination of blood pressure in fetal life, (2) long-term 'programming' of lipid metabolism through feeding during infancy, and (3) the early setting of haemostatic mechanisms.

To study the effect of maternal physique and intrauterine growth on adult blood pressure we traced 449 men and women born in a hospital in Preston,

TABLE 2   Standardized mortality ratios for ischaemic heart disease according to birth weight and weight at one year in men who were breast fed

| Weight at one year (lb) | Weight at birth (lb) | | | |
	Below average (≤7)	Average 7.5–8.5	Above average (≥9)	All
Below average (≤21)	100 (80)	100 (77)	58 (17)	93 (174)
Average (22–23)	86 (34)	87 (67)	80 (29)	85 (130)
Above average (≥24)	53 (14)	65 (42)	59 (32)	60 (88)
All	88 (128)	85 (186)	65 (78)	81 (392)

Numbers of deaths in parentheses.
From Barker et al 1989b.

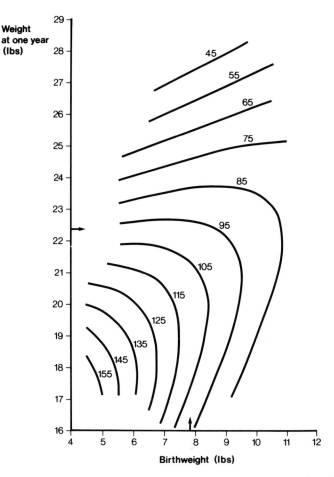

FIG. 2.   Relative risk for ischaemic heart disease in men who were breast fed according to birth weight and weight at one year. Lines join points with equal risk. Arrows indicate mean weights. (From Barker et al 1989b by permission of the Editor of *The Lancet*.)

Lancashire during 1935–1943 and measured their blood pressures (Barker et al 1990, Barker 1990). The birth records in the hospital were unusually complete, including, for example, seven measurements of the infant head. We found that the blood pressure and risk of hypertension among men and women aged around 50 years was strongly predicted by a combination of placental and birth weight (Table 3). Systolic and diastolic pressures rose as placental weight increased and fell as birth weight increased. These relations were independent, the highest pressures occurring among people who had been small babies with large placentas. Higher body mass index and alcohol consumption were also associated with higher blood pressure, in keeping with the results of many other studies

TABLE 3   Mean systolic pressure (mmHg) of 449 men and women aged 46 to 54 years according to placental weight and birth weight

Birth weight (lb)	Placental weight (lb)				
	-1.0	-1.25	-1.5	>1.5	All
-5.5	152	154	153	206	154 (45)
-6.5	147	151	150	166	151 (106)
-7.5	144	148	145	160	149 (169)
>7.5	133	148	147	154	149 (129)
All	147 (68)	149 (171)	147 (120)	157 (90)	150 (449)

Numbers of people in parentheses.
From Barker et al 1990.

(Intersalt Co-operative Research Group 1988), but the relation of placental and birth weight to blood pressure levels, and to established hypertension, was independent of these influences and stronger.

Our data point to a possible mechanism for the relation between placental weight and blood pressure. Studies of fetal blood flow in animals have shown that in response to hypoxia there is a redistribution of fetal cardiac output which favours the perfusion of the brain (Campbell et al 1967, Rudolph 1984). Professor K. L. Thornburg will be describing this phenomenon in detail (1991: this volume). In our data, greater placental weight at any birth weight was associated with a decrease in the ratio of length to head circumference. This disproportionate growth is consistent with diversion of blood away from the trunk in favour of the brain. A fetal circulatory change of this kind, occurring in a fetus that is small in relation to its placenta, could be associated with irreversible consequences, perhaps by changes in arterial structure. There is evidence in animals and humans that changes in blood flow in early life can alter arterial structure and compliance (Berry & Greenwald 1976, Meyer & Lind 1974, Berry et al 1976).

These findings raise the question of what environmental influences act on the mother and determine placental and birth weight. In particular, what determines the discordance between placental and fetal size which leads to high blood pressure? Little is known about this. We suspect that maternal physique and nutrition are the key influences. But at present our conclusion is simply that environmental influences acting in fetal life have a major effect on adult blood pressure and hypertension.

In collaboration with Professor C. N. Hales, of the Department of Biochemistry, University of Cambridge, we are currently examining lipid levels in a sample of men born in Hertfordshire during 1911–1930. Table 4 shows some early results. Among 108 men, all of whom were breast fed, total

TABLE 4 Blood cholesterol and fibrinogen concentrations in men aged 65 years, who were breast fed, according to weight at one year

Weight at one year (lb)	Cholesterol (nmol/l)	Fibrinogen (g/l)
≤20	6.9 (16)	3.12 (25)
−22	6.3 (37)	3.07 (47)
−24	6.2 (37)	3.08 (47)
≥25	6.0 (18)	2.96 (27)

Numbers of men in parentheses.

cholesterol levels were inversely related to weight at one year. This is consistent with the higher risk of ischaemic heart disease in men who were lighter at birth and at one year. As yet, the numbers of men who were bottle fed is too small for analysis. We shall shortly have results for around 500 men. Our tentative conclusion from these early results is that nutrition and growth in fetal and infant life affect adult lipid metabolism. Dr G. E. Mott will be describing experiments which suggest that infant feeding in baboons programmes cholesterol metabolism in the adult (Mott et al 1991: this volume).

Finally, in collaboration with Dr T. W. Meade, we are examining levels of fibrinogen and Factor VII in men in Hertfordshire. Fibrinogen and Factor VII are strongly associated with the risk of ischaemic heart disease (Meade et al 1986, Meade & North 1977). Table 4 shows early results. Among 146 men, fibrinogen levels are inversely related to weight at one year.

In conclusion, detailed geographical analyses in England and Wales suggest that poor maternal physique and nutrition, and poor fetal and infant growth, are associated with increased risk of cardiovascular disease in adult life. In a follow-up study of men born around 70 years ago, who were breast fed, those with the lowest weights at birth and one year had the highest death rates from ischaemic heart disease. Follow-up and examination of men and women who are still alive has shown strong relations between early growth and three major risk factors for cardiovascular disease: high blood pressure, high cholesterol and high fibrinogen. Maternal, fetal and infant influences seem much more important in the causation of cardiovascular disease than we have previously supposed.

References

Barker DJP 1990 The intrauterine origins of adult hypertension. In: Dawes GS (ed) Fetal autonomy and adaptation. Wiley, Chichester
Barker DJP, Osmond C 1986 Infant mortality, childhood nutrition, and ischaemic heart disease in England and Wales. Lancet 1:1077–1081
Barker DJP, Osmond C 1987 Death rates from stroke in England and Wales predicted from past maternal mortality. Br Med J 295:83–86

Barker DJP, Osmond C, Law C 1989a The intra-uterine and early postnatal origins of cardiovascular disease and chronic bronchitis. J Epidemiol Community Health 43:237–240

Barker DJP, Winter PD, Osmond C, Margetts B, Simmonds SJ 1989b Weight in infancy and death from ischaemic heart disease. Lancet 2:577–580

Barker DJP, Bull AR, Osmond C, Simmonds SJ 1990 Fetal and placental size and risk of hypertension in adult life. Br Med J 301:259–262

Berry CL, Greenwald SE 1976 Effects of hypertension on the static mechanical properties and chemical composition of the rat aorta. Cardiovasc Res 10:437–451

Berry CL, Gosling RG, Laogun AA, Bryan E 1976 Anomalous iliac compliance in children with a single umbilical artery. Br Heart J 38:510–515

Campbell AGM, Dawes GS, Fishman AP, Hyman AI 1967 Regional redistribution of blood flow in the mature fetal lamb. Circ Res 21:229–235

Campbell JM, Cameron D, Jones DM 1932 High maternal mortality in certain areas. (Ministry of Health Reports on Public Health and Medical Subjects, No. 68) HMSO, London

Gardner MJ, Crawford MD, Morris JN 1969 Patterns of mortality in middle and early old age in the county boroughs of England and Wales. Br J Prev Soc Med 23:133–140

General Register Office 1911 et seq. Registrar General's statistical reviews of England and Wales, 1911 et seq. Part I. Tables, medical. HMSO, London

Intersalt Co-operative Research Group 1988 Intersalt: an international study of electrolyte excretion and blood pressure. Results for 24 hour urinary sodium and potassium excretion. Br Med J 297:319–328

Local Government Board 1910 Thirty-ninth annual report 1909–10. Supplement on infant and child mortality. HMSO, London

Meade TW, North WRS 1977 Population-based distributions of haemostatic variables. Br Med Bull 33:283–288

Meade TW, Mellows S, Brozovic M et al 1986 Haemostatic function and ischaemic heart disease: principal results of the Northwick Park Heart Study. Lancet 2:533–537

Meyer WW, Lind J 1974 Iliac arteries in children with a single umbilical artery: structure, calcification and early atherosclerotic lesions. Arch Dis Child 49:671–679

Mott GE, Lewis DS, McGill HC Jr 1991 Programming of cholesterol metabolism by breast or fomula feeding. In: The childhood environment and adult disease. Wiley, Chichester (Ciba Found Symp 156) p 56–76

Registrar General 1978 Registrar General's decennial supplement: occupational mortality in England and Wales 1970–72. HM Stationery Office, London

Rudolph AM 1984 The fetal circulation and its response to stress. J Dev Physiol 6:11–19

Thornburg KL 1991 Fetal response to uterine stress. In: The childhood environment and adult disease. Wiley, Chichester (Ciba Found Symp 156) p 17–37

DISCUSSION

Hamosh: Professor Barker, can one dissect out the nutritional effects on later cardiovascular disease in the offspring, both of the mother's nutrition and, even better, the grandmother's nutrition, from the environmental effects of toxins, xenobiotics, and so on? And could one also examine separately the effects of *in utero* maternal malnutrition and/or infant malnutrition (during the first year of life) from later nutrition? For example, if the poorest group, with the poorest predictors, had been transferred after three months or one year of life to the most optimal environment, what would the outcome be in terms of later cardiovascular disease?

Barker: My view is that the evidence does not point to environmental toxins as having important effects on cardiovascular diease. Rather, the evidence points to effects of adverse nutrition of women, going through several generations. Certainly my intention is to focus our research on the nutrition of the mother, and on her physique, which partly depends on her nutrition as a young girl. What we know from a large study of migrants in England and Wales is that people born in areas of low risk of cardiovascular disease—London, for example—carry with them part of their low risk wherever they migrate to within the country. I do not think that the uncertain inferences from migrant studies will tell us the relative importance of early and late nutrition. Rather, understanding seems likely to come from further long-term studies of individuals, and from an understanding of underlying mechanisms.

Hanson: You are arguing that the small, undernourished mother produces a small fetus, and also that she has a relatively large placenta, in relation to fetal weight. You are therefore suggesting that there is some link between maternal nutritional status and placental growth. I would go along with that on the grounds that we don't actually know what controls placental growth. Is this link there in your data?

Barker: It's not there in my data. I suspect that in larger placental size you are seeing a marker of poor maternal nutrition.

Thornburg: It is well known that there is a direct relationship between fetal size and placental size (or placental blood volume) in mammals in general (Owens & Robinson 1988) and particularly in people (Bonds et al 1984). If that relationship is also generally true in the UK, it means that you are basing your relationships on those individuals who don't fit the apparently normal pattern.

Barker: Yes. A lot is known about babies in whom the placenta is relatively small. We are now seeing the other kind of disproportion, where the placenta is too big; and this is new.

Dobbing: The retrospective approach obviously has the usual limitations. Speaking of low birth weight babies in 1921–1922, the low birth weight group is compounded of those born too soon (prematures) and those born too small but of appropriate gestational age. It may be known whether, in 1921–1922, low birth weight babies were more predominantly prematurely born than now?

Barker: More of them were premature than now, but my understanding of the literature at that time is that most small babies were the result of intrauterine growth retardation, not prematurity.

Dobbing: Whichever way it was, it would make a difference to one's further thinking about mechanisms for the phenomena you are suggesting. Also, if the low birth weight babies were predominantly small-for-dates babies, it would again be important to know what variety of small-for-dates they were, because only a proportion of them are thought to be due to what might loosely be called 'fetal malnutrition'.

Barker: Yes. In the Preston study, the length of gestation of most of the subjects was known. It was unrelated to subsequent blood pressure. There is therefore no evidence that premature expulsion from the uterus is a risk factor for high blood pressure. Secondly, the characteristic of the babies whose placental weight is too great in relation to their birth weight is that their head size was big in relation to their length.

Dobbing: So they were disproportional small-for-dates babies? If so, I think that is an important aspect which should be stated, quoting the supporting evidence. Again, I wonder whether the range of variation in placental weight might be partly due to blood content and, if that is so, whether something about the physiology of the expulsion of blood from the placenta before it was weighed might have a bearing on this mechanism.

Hanson: You are suggesting that the weight of the placenta may be affected by whether or not the blood has been squeezed out before the placenta is weighed?

Dobbing: Yes, either squeezed by the midwife, or squeezed by the placental vessels themselves, which would be more interesting. It is conceivable that the vascular properties of some placentas squeeze the blood out more, and others less so, which would account for that difference in weight—a difference that is not great in relation to the blood content of the average placenta.

Hanson: Certainly the fascinating point about David Barker's findings is that one is not dealing with very large placentas or with very small babies; both are within the normal range.

Lloyd: Do you really have those data? You simply specified babies 'under 5.5 lb'; do you know how many were very much smaller?

Barker: The smallest baby was 3 lb 2 oz and the next two were 4 lb.

Lloyd: I am interested in the maternal weights, because you talked about maternal undernutrition. Do you actually have any data on maternal nutrition, or is it just supposition that these were undernourished mothers?

Barker: The only data in Preston are on maternal pelvic size, which correlated strongly with birth weight, as one would expect. In another set of data in Sheffield which we discovered subsequently, and are now doing the same study on, we do have maternal weights.

Chandra: One situation where the birth weight is low and placental weight relatively high is intrauterine infection. Such infections are more common in women from the lower socio-economic levels or in those who are undernourished.

Casaer: Was the distribution of the placental weights by any chance bimodal?

Barker: We don't see this. The distribution of placental weight approximates to a normal distribution skewed to the right. It's not bimodal.

Chandra: Are the subjects you have analysed in Hertfordshire those who had *not* migrated out of the area?

Barker: Out of 7991 boys born in the study period, 5654 were either living in England and Wales, or had died.

Chandra: Another question relates to the recording of birth weight, or weight at one year. Those of us working in these areas feel that it needs a lot of training of the individuals who weigh infants, and very accurate scales, to achieve the minimum possible intra-observer or inter-observer variance. Since these infants in rural Hertfordshire in the 1911–1930 period were weighed on potato scales, and when one weighs potatoes one often rounds them up the nearest half pound or pound, would that have made any big difference to the overall assessment?

Barker: The crudeness of the measurements is in stark contrast to the precision of the predictions. Presumably, these predictions would have been better still if the weights had been measured better!

Richards: Taking up the issue of migration, height is correlated with social class, and assortative mating (the tendency of people to marry people with similar characteristics to themselves) is strong for height. Social mobility is also related to height, for both men and women, with taller individuals more likely to be upwardly mobile. This ought to mean that, if height is a rough indicator of maternal physique, there may be a selective effect whereby populations in the poorer rural and industrial areas will lose, through migration, their fitter and taller members. Women moving out of such areas may tend to be those who are taller and are likely to produce bigger babies. This may exaggerate the effect that you show in your geographical distributions. This may be relevant to the question of London, where historically there has been a strong inward migration of, presumably, the healthier women coming from the poorer areas, as well as the healthy men.

Barker: That is an extremely helpful comment, which fits in with what we know occurred, from Charles Booth's survey (*Life and Labour in London*), namely that there was a constant renewal of London by immigration of the fittest young women from an area which started in the west, in Devon, and extended up to Norfolk. They mostly came for one reason, domestic service, in which employment they continued to be well nourished. The picture was of London sucking in generations of the best young women, who had the lowest mortality in childbirth, and whose babies had the lowest neonatal mortality. Thereafter, their children had a poor environment and, as Booth wrote, 'after two generations London life reduces the immigrants' descendants to the level of those among whom they live'.

Hamosh: Do you have data on the siblings in your Hertfordshire study? I ask this because I would like to know how long the breast-feeding period was at that time. Could it be that maternal reserves were depleted by the time a second baby was conceived and born?

Barker: I have that information, because we have data on all the children born in Hertfordshire over a 35-year period. We have not yet looked at the effect of parity.

Lucas: The increase in vascular diseases is very recent and has occurred in only some Western countries, but presumably adverse perinatal and childhood factors, maternal malnutrition, variations in placental weight, and the other factors you talked about, would have existed in previous centuries and in other cultures where the incidence of vascular disease was very low. How do you fit that into your scheme?

Barker: The ecological data suggest that stroke mortality is related principally to events in intrauterine life. The suggestion that is emerging is that blood pressure is 'set' during the intrauterine period. Stroke mortality has declined every year for the past 40 years in Britain and in many other countries, which is consistent with past improvements in maternal physique and health. By contrast, ischaemic heart disease mortality has risen very steeply in industrialized countries, so there is something else, perhaps involving cholesterol metabolism or clotting factors, that is set in early life. To explain the distribution of ischaemic heart disease it is necessary to postulate two sets of factors: one related to poor living standards, which operates in very early life, and another set relating to good living standards which operates later, and presumably is associated with the high energy Western diet.

Mott: Your follow-up of individuals still living, in the Hertfordshire study, is extremely important. Measuring lipoprotein cholesterol would be valuable in addition to total serum cholesterol, because, at least in the non-human primate, programming effects of early diet act primarily upon the lipoproteins and not on the total serum cholesterol.

Barker: We are indeed measuring lipoprotein in people born in Hertfordshire.

Wood: Rather naively, but hopefully, I am wondering whether our present feeding practices for newborns are also going to reduce the problem of cerebro-vascular and ischaemic heart disease further in 30 years' time. In the UK we now have better-fed babies, who are fed on demand. We still have low birth weight babies, but we have very much lower neonatal mortality than in the period you are discussing. In addition, we have another small cohort of very small preterm babies who are preserved in the face of immense difficulty. One would like to think that our improved neonatal practices are likely to influence outcome in terms of blood pressure and cardiovascular disease; but if the setting of, say, peripheral resistance is predetermined, before the fetus gets near to the time of delivery, such expectations and aspirations may be over-optimistic.

Barker: The framework of ideas within which we are working is that circulatory adaptations in the fetus lead to changes in arterial structure, with reduced compliance. This in turn leads to higher pulse pressure, and to further charges in arterial structure. This feedback could perpetuate high levels of systolic pressure from infancy to old age.

Thornburg: In your Fig. 1 (p 4) you plotted cardiovascular mortality at ages 55 to 74 as a function of infant mortality. I was interested in the outliers. There were occasional points where the infant mortality seemed to be rather high, yet the death from cardiovascular disease was rather low. Can you learn anything about the outliers? Do they fit in with your hypothesis on nutrition, for example?

Barker: Unfortunately, the outliers are simply places that are extremely small, such as the City of London, which was a London borough with few inhabitants. Mortality rates in these places are liable to wider fluctuations.

Thornburg: So the answer is that you can't learn from these points? Can you learn anything from the Dutch study of the Hunger Winter of 1944 to 1945 (Stein & Susser 1975), when many people in Holland starved?

Barker: Yes. The offspring, who were *in utero* at that time, have been followed up. People who were *in utero* in the last trimester of pregnancy during that winter, although born small, were of normal height and had normal intellectual development at age 18. More recently, Dr Lumey has studied those who were *in utero* in the first trimester at that time; they were born with normal weight and at 18 had normal height, but when the women had babies, their babies were smaller than predicted (Lumey 1988).

Suomi: In the United States today there is a relatively large cohort of babies being born to extremely small mothers, and they are typically born unusually small. This is not so much because of the other factors mentioned, such as prenatal diet, but because the mothers are very young, 13–15 years of age. Do you have any thoughts on the degree to which your findings might generalize to this population?

Barker: In the UK, women having children young was a feature of coal-mining communities, who have conspicuously bad health. However, there were few very young mothers in the Preston study and we do not know mothers' ages in Hertford. So I cannot generalize here.

Dobbing: You didn't tell us whether it was better to be breast fed or artificially fed, in the Hertfordshire study. I suppose it can be assumed that qualitatively the breast milk in those days, in the earlier part of this century, was not much different qualitatively from nowadays, although quantitatively it may have been; whereas the artificial food is likely to have been very different from now. Apart from that, is it better to have been breast fed in 1922?

Barker: Although, on the face of it, it was better to have been breast fed in Hertfordshire, in terms of the overall lower mortality rates from ischaemic heart disease, we don't yet have sufficient numbers to take account of the fact that the babies who were bottle fed were different; they were, on average, smaller at birth, and one needs to allow for that. In the Hertfordshire study, the information that I long to be able to give Glen Mott is whether the cholesterol levels are higher in the breast-fed or the bottle-fed people, taking account of birth and infant weight.

Dobbing: Yet the breast-fed individuals would have had higher cholesterol levels in infancy.

Hamosh: But that could be a transient phenomenon, during the day, with breast feeding.

Dobbing: We are talking of transient phenomena.

Richards: This very small minority of mothers who chose to bottle feed in the 1920s would be nothing like a bottle-feeding group today. Choice of feeding method has varied greatly over time, so the social composition of a group choosing a feeding method will be specific to a particular historical period. In so far as social factors are related to maternal health, there may also be differences in the growth and health of babies produced by mothers who opt for each feeding method.

Barker: There's every reason to believe that that is true. We know quite a lot about the kind of artificial infant feeding preparations available in 1920. They were extremely varied in their compositon. They were widely used in Lancashire cotton towns, where women returned to work soon after delivery, leaving their babies in the care of child minders. In places like Hertfordshire, artificial feeds were not widely used. It was a community with a tradition of breast feeding, and with a 200-year history of being a wet-nursing area for babies brought out from London. There wasn't much cows' milk available because it was mostly sold to London. So the minority of mothers who did not breast feed must have been unusual. What they fed their babies with, we cannot tell.

Casaer: Were they not the infants whose mothers died?

Barker: No. It wasn't that; we know that much.

Murray: There were striking seasonal variations in infant mortality at the time of the First World War, and earlier. Have you looked at the effect of season of birth in your study?

Barker: There were seasonal variations in post-neonatal mortality (that is, deaths from one month to one year). These variations were due to higher rates for respiratory mortality in the winter and for diarrhoeal disease in the summer, but the relationships that we are looking at are primarily with *neonatal* mortality.

Murray: Still births also showed a variation, but not to such an extent.

References

Bonds DR, Gabbe SG, Kumar S, Taylor T 1984 Fetal weight/placental weight ratio and perinatal outcome. Am J Obstet Gynecol 149:195–200

Hamosh M 1988 Does infant nutrition affect adiposity and cholesterol levels in the adult? J Pediatr Gastroenterol Nutr 7:10–16

Lumey LH 1988 Obstetric performance of women after in utero exposure to the Dutch famine (1944–45). PhD thesis, Columbia University, NY, USA

Owens JA, Robinson JS 1988 The effect of experimental manipulation of placental growth and development. In: Cockburn F (ed) Fetal and neonatal growth. Wiley, Chichester, p 49–77

Stein Z, Susser M 1975 The Dutch famine 1944/45 and the reproductive process. I. Effects on six indices at birth. Pediatr Res 9:70

Fetal response to intrauterine stress

Kent L. Thornburg

Department of Physiology, School of Medicine, Oregon Health Sciences University, Portland, Oregon 97201, USA

Abstract. Many human infants are born inappropriately small as a result of stress suffered during intrauterine life. Acute reductions in oxygen delivery to fetal tissues have therefore been studied in animals so that insight can be obtained into the adaptive mechanisms that underlie human developmental abnormalities. It is now known that during moderate hypoxic stress fetal arterial blood pressure is variably increased while heart rate and cardiac output are depressed; blood volume is reduced but cardiac output is redistributed to spare the myocardium, brain and adrenal glands at the expense of most other organs. Also a greater fraction of oxygen-rich venous blood from the placenta is returned to the heart for distribution. Spared organs are those that grow disproportionately well in human asymmetrical intrauterine growth retardation (IUGR). These cardiovascular responses are not fully understood although elevated fetal plasma levels of catecholamines and a host of fetal hormones are undoubtedly important. Chemical sympathectomy does not abolish the blood flow redistribution phenomenon, which implies that autoregulatory effects may be responsible for some of the redistribution of blood flow. Fetal hypoxaemia and metabolic abnormalities are sequelae often found with human IUGR, suggesting placental exchange defects. IUGR placentas appear to have defective transport mechanisms for many nutrients. Animal studies suggest that the placenta will give priority to its own needs over those of the fetus, when necessary, to support its own growth and function.

1991 The childhood environment and adult disease. Wiley, Chichester (Ciba Foundation Symposium 156) p 17–37

This chapter gives a brief overview of fetal responses to several well-studied types of hypoxic stress. This topic was addressed in a previous review (Rudolph 1984). The relationship between intrauterine growth retardation (IUGR) and fetal hypoxic stress is unclear because the complexity of IUGR has so far prevented a clear understanding of any of its many aetiologies. Aside from the genetic causes of retarded growth, IUGR is also ascribed to 'environmental' causes (Warshaw 1986), among which is uteroplacental insufficiency. The latter term is a catch-all phrase used to describe conditions that inhibit the generous feto-maternal exchange of nutrients, gases and water, when due to insufficient placental perfusion or placental structural defects. Unfortunately, the nature of such defects in flow or structure has not been well described (in fact, some investigators are sceptical whether true placental defects ever exist). Furthermore, it should be noted that even a close correlation between the degree of fetal growth retardation and indices of fetal distress does not necessarily point to the origin of the circular cause-and-effect relationship. Nevertheless, the motivation underlying much of the clinical and animal research on defective growth has been to learn more of the fetal adaptations to 'quantifiable' hypoxic stress with the hope that the underlying pathophysiology of intrauterine growth abnormalities can be discovered.

Normal distribution of fetal cardiac output

The normal fetal circulatory pattern is different from that of the adult; fetal shunts (foramen ovale, ductus arteriosus) allow the fetal right and left ventricles to pump in parallel rather than in series (Dawes 1968). Therefore, cardiac output in the fetus is defined as the sum of the stroke volumes of the combined ventricles per minute. In sheep, about 40% of the cardiac output is directed to the placenta, where feto-maternal gas exchange takes place (Anderson et al 1981). Oxygenated placental blood, which flows back to the heart through the umbilical vein, the ductus venosus and the inferior vena cava, is diverted by the foramen ovale into the left side of the heart. Well-oxygenated blood from the left ventricle then supplies the heart and brain.

The use of radiolabelled microspheres has allowed the careful measurement of fetal cardiac output and its distribution. The microsphere technique works on the principle that millions of small plastic radiolabelled spheres can be injected through catheters into the heart where they mix and distribute to the organs of the body in proportion to the blood flows of those organs. Because the spheres are inflexible and about twice the diameter of an ordinary red cell (15 μm), they become trapped in the microcirculation of the organs into which they flow. The actual flow to any organ is then determined by counting the radioactivity of the organ in question and comparing the counts from a reference sample drawn at a known flow rate during the experiment. For example, if the reference blood sample is drawn at 10 ml per minute and contains spheres yielding 100 counts/min, then each 100 counts/min in a particular organ represents a flow of 10 ml/min.

In a monumental study, Rudolph & Heymann (1970) calculated the blood flows for individual organs in sheep fetuses of different gestational ages using the radiolabelled microsphere technique. (We are also in debt to Professor Rudolph and colleagues for many of the data on the fetal flow distributions under various conditions of stress, as discussed below.) They determined that the cardiac output of the sheep fetus is about 500 ml/min per kg fetal body weight, of which at least 60% is ejected from the right ventricle. Of the total cardiac output, about 4% is delivered to brain, 3% to heart, 5% to gut, 2% to kidneys, and 40% to the remaining fetal carcass. These values apply to the near-term fetal lamb (>125 days out of the 150 days of gestation). Ordinarily, the well-oxygenated umbilical venous blood leaving the placenta has a P_{O2} of about 30 mmHg and a P_{CO2} of about 38 mmHg, whereas arterial blood taken from the descending aorta has a P_{O2} of about 25 mmHg and a P_{CO2} of 45 mmHg. Fetal arterial pH is normally about 7.35.

Animal models of fetal hypoxic stress

A number of animal models have been developed with which to determine fetal growth responses to intrauterine stress (Owens & Robinson 1988). Because maternal circulatory abnormalities are believed to result in fetal hypoxic stress and abnormal fetal growth, various methods of reducing oxygen delivery to the fetus have been devised. Four different pregnant sheep models have been particularly useful: (1) the reduction of the fraction of inspired oxygen to the ewe, (2) fetal haemorrhage, (3) compression of the umbilical cord and (4) the restriction of uterine artery blood flow.

One of the problems in such models is standardizing the level of stress produced and applying these levels to clinical situations. When considering these models it is important to note the level of stress chosen by the investigator for any particular experiment, because that level is crucial to the interpretation and application of the data.

Reduced partial pressure of maternally inspired oxygen

Maternal hypoxaemia and the resulting fetal hypoxaemia are induced by placing a large bag around the head of a pregnant ewe so that the inspired oxygen fraction of the ewe can be regulated. (Many investigators now infuse nitrogen directly into the maternal trachea via a small catheter, allowing the dilution of alveolar gas to any desired level.) Cohn et al (1974) measured the distribution of fetal cardiac output under hypoxaemic conditions in unanaesthetized sheep. Pregnant ewes were given a gas mixture designed to produce a consistent fall in fetal arterial P_{O2} to around 12 mmHg without changing fetal CO_2 levels. During the hypoxaemic period, fetal arterial pressure rose slightly and heart rate decreased by 25%. Figure 1 shows the changes in blood flow to selected organs. Actual oxygen deliveries (the product of blood flow and oxygen content) to selected organs were not reported by Cohn et al but they have been estimated here from the organ flow–O_2 content relationship (Longo et al 1976, Peeters et al 1979). Estimated oxygen deliveries are included in Fig. 1 only to indicate whether flow alterations are adequate to maintain the flow of oxygen to any given organ and should be considered as rough approximations only. Contrary to the report of Rudolph (1984), the dramatic increases in oxygen delivery to heart and brain reported by Reuss & Rudolph (1980) are not applicable here because they represent umbilical vein-derived oxygen delivery only and not total organ oxygen delivery.

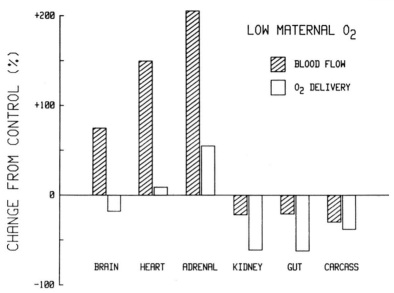

FIG. 1.   Changes in blood flow and estimated oxygen deliveries to selected sheep fetal
organs during an acute fetal hypoxaemic episode induced by allowing the ewe to breath air
containing a reduced fraction of oxygen (Cohn et al 1974). Oxygen deliveries were estimated
from published relationships between organ flows and blood oxygen contents for various
organs and the estimated contents from similar experiments in other laboratories (Longo
et al 1976, Peeters et al 1979); oxygen deliveries are only rough approximations.

   During the period of hypoxaemia, Cohn et al (1974) found that blood flow
to heart, brain and adrenal glands increased dramatically. This increase in blood
flow did not quite maintain estimated oxygen delivery to the brain. A similar
result was found by Peeters et al (1979) and Ashwal et al (1980), though both
groups demonstrated that oxygen delivery to the brainstem portion of the
CNS was maintained. Oxygen delivery to the heart and adrenals was actually
greater than under control conditions. Under these hypoxic conditions, the heart
received 7% of the cardiac output and the brain 6%, rather than their usual
3–4%. This increased flow was made at the expense of other organs. Blood
flow to the kidneys, the gut and carcass was reduced significantly and, because
the oxygen content of the blood being delivered to these organs was also reduced,
so was oxygen delivery. Pulmonary and splenic blood flows were somewhat
reduced also.

Reduction of fetal blood volume by haemorrhage

Itskovitz et al (1982) determined the redistribution of fetal cardiac output after
reducing the fetal blood volume by controlled haemorrhage. Organ blood flows

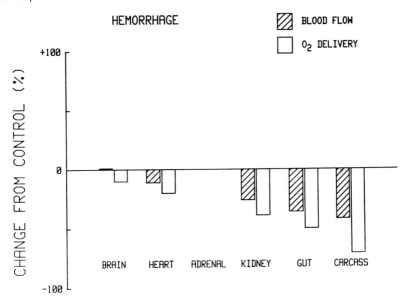

FIG. 2.   Changes in blood flow and oxygen deliveries to selected fetal organs after loss of fetal blood volume by fetal haemorrhage (approximately 20% blood loss). Note that unlike in Fig. 1, brain and heart both show small decreases in oxygen delivery following haemorrhage (see Itskovitz et al 1982).

were determined 5–10 min after 22% of the estimated fetal blood volume had been removed (this would be some 100 ml from a 4 kg sheep fetus). Under these conditions, fetal cardiac output was reduced by 30% while heart rate and arterial blood pressure did not change. Even though umbilical blood flow dropped by more than 20% and haematocrit was reduced from 32 to 28%, aortic blood gases did not change. Oxygen consumption by the fetus was maintained.

Figure 2 shows the alterations in organ blood flow and oxygen delivery after haemorrhage. Blood flow to the brain did not change, though oxygen delivery appeared to be slightly less than normal. The heart suffered a slight loss of flow and oxygen delivery (with its apparent reduced work load). Yet the brain and heart received enhanced portions (50 and 35% increases, respectively) of the cardiac output. Figure 2 shows that the changes in blood flows and oxygen deliveries for the kidneys, gut and carcass were remarkably similar to those seen for reduced maternal arterial P_{O2} (Fig. 1); blood flow was reduced by 25–50% with oxygen deliveries reduced even more.

Reduction of umbilical blood flow by cord constriction

After the above study of fetal haemorrhage, Itskovitz et al (1987) published a second study of fetal stress. In these experiments an inflatable occluder was placed around the umbilical cord of nine fetal sheep of >120 days gestation. Organ flows were measured before and during reductions of umbilical flow by 50%, which caused a large increase in upstream mean umbilical venous pressure (32 mmHg, up from 12 mmHg), a 20% reduction in cardiac output, a 15% decrease in heart rate and a 10% decrease in mean arterial pressure. Upper body arterial P_{O2} dropped by 23% to 18 mmHg and descending aortic blood P_{O2} dropped by a similar proportion to 16 mmHg. Figure 3 shows the flow redistribution. As with the maternal hypoxaemia experiments, fetal blood flows to brain, heart and adrenals were significantly increased with cord constriction. These increases in flow were significant in maintaining oxygen delivery to brain and heart. The flow response of other organs was different from that in the hypoxaemia studies, however. The oxygen deliveries to kidney, gut and carcass were somewhat decreased, even though flows to kidney and carcass were actually increased.

Reduction of uterine blood flow by uterine artery constriction

Many methods have been used to reduce maternal blood flow to the uterus (Clark & Skillman 1984). The adjustable Teflon vascular clamp method was used by

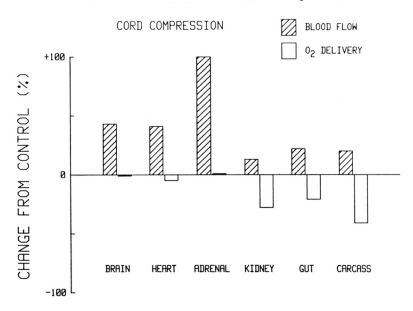

FIG. 3.   Changes in blood flow and oxygen deliveries to selected fetal organs during a 50% reduction in umbilical blood flow caused by umbilical cord occlusion with an inflatable occluder. Note large increases in blood flow to heart, brain and adrenal gland preventing loss of oxygen flow to those organs at the expense of other organs (Itskovitz et al 1987).

Bocking et al (1988) to study the responses to reduced uterine blood flow in 128-day sheep fetuses. Blood flow distributions in the fetus were determined with microspheres before and after uterine blood flow was reduced to a level causing a 40% decrease in fetal arterial oxygen saturation (Fig. 4). Fetal flow measurements were then made at one hour, 24 hours and 48 hours after uterine artery constriction. After one hour of reduced uterine blood flow the fetal arterial P_{O2} decreased from 23 mmHg to 17 mmHg; P_{CO2} rose to 56 mmHg and pH dropped to 7.23. Fetal heart rate and blood pressure increased insignificantly after a transient bradycardia. Unfortunately, neither cardiac output nor oxygen delivery to individual organs was reported. However, as for studies in the first group, oxygen deliveries were estimated (complicated further because it was not always clear which arteries were used for blood gas samples). These rough estimates are included in Fig. 4, along with the accurately obtained organ flows.

The fetus responded in a pattern that has become, by now, predictable. Blood flows to brain, heart and adrenal glands more than doubled in response to reduced oxygen contents and elevated CO_2 tensions. These increased flows

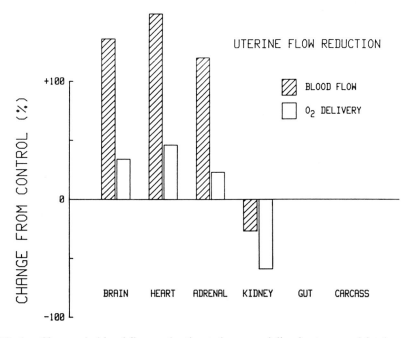

FIG. 4.   Changes in blood flow and estimated oxygen deliveries to several fetal organs during reduction in uterine blood flow produced by a mechanical occluder on the uterine artery (Bocking et al 1988). Note substantial increase in blood flow to brain, heart and adrenal gland which preserved (estimated) oxygen flow to those organs. Oxygen deliveries were estimated as in Fig. 1. Blood flows to gut and carcass were not measured in this study; however, nuchal muscle blood flow was measured and did not change.

were apparently adequate to maintain oxygen deliveries. Kidney blood flow and its associated estimated oxygen delivery were somewhat reduced. As with previous experiments, heart and brain were apparently 'protected' from the ravages of hypoxia, while most other organs were not.

Control of the venous shunts during hypoxic stress

Since the placenta is the source of oxygen for the fetus, it would seem prudent to have a mechanism for directing highly oxygenated placental blood away from non-vital organs and toward the brain and myocardium in times of stress. It appears that such a mechanism is in operation under certain conditions. During the episodes of fetal hypoxaemia resulting from reduction of maternal arterial oxygen saturation (Reuss & Rudolph 1980), following fetal haemorrhage (Itskovitz et al 1982) or during cord compression (Itskovitz et al 1987), the portion of venous return from the placenta that bypassed the liver and flowed directly to the inferior vena cava was increased. In the hypoxaemia experiments this altered the venous return pattern so that nearly 40% of the cardiac output was 'placental blood', up from the control value of 27%.

Mechanisms that underlie fetal responses to oxygen deprivation

It is clear from these models that there is a general response to oxygen deprivation stress in the fetus. Cardiac output is depressed. Blood flows to brain, heart and adrenal gland are maintained or augmented even if blood flow deprivation for other tissues is required. During fetal hypoxaemia, arterial pressure may rise and heart rate is slowed. Venous blood flow patterns are altered. The mechanisms that underlie these complex physiological responses are not at all well defined. But several mechanisms may be proposed as key ingredients in the response to hypoxaemia. These might include baroreceptor and chemoreceptor reflexes, and autoregulation, in addition to the direct release of catecholamines by the adrenal medulla, or the release of vasopressin, renin, and/or atrial natriuretic peptide, among others.

Autonomic control

In 1983, Iwamoto et al administered 6-hydroxydopamine to lambs ranging in gestational age from 115 to 138 days. 6-Hydroxydopamine, a sympathetic neurotoxin that destroys sympathetic nerve terminals, causes a chemical sympathectomy when administered to animals. When fetuses so treated were subjected to hypoxaemia by depressing the maternal inspired oxygen levels, cardiac output was preserved while arterial pressure rose slowly. Blood flow to heart, brain and adrenals increased dramatically. Lung blood flow was reduced, gut flow increased, and flows to kidney and carcass were unchanged.

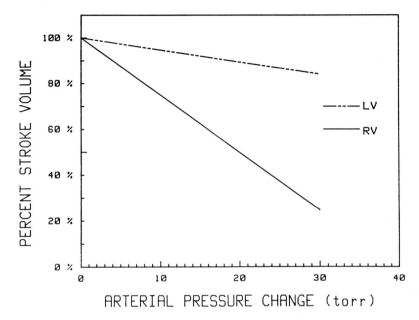

FIG. 5.   Changes in right and left ventricular stroke volume (% control) as a function of simultaneous increases in blood pressure in the pulmonary artery and ascending aorta of fetal sheep. Pressures were increased by inflating an occluder around the postductal descending aorta. Note that right ventricular (RV) stroke volume is severely reduced by a 20 mmHg increase in pressure whereas the left ventricle (LV) is only mildly affected. (From Reller et al 1987 with permission of *Pediatric Research*.)

In short, the response in the chemically sympathectomized fetus was remarkably similar to responses in intact fetuses, leaving the role of the autonomic nervous system open to question.

However, there is little doubt that adrenally derived circulating catecholamines are important in the fetal response to hypoxaemia. There is mounting evidence for the following scenario: the adrenal gland is directly stimulated by falling partial pressures of oxygen to release catecholamines. Adrenaline and noradrenaline stimulate arteriolar α-adrenergic receptors, causing an increase in general body vascular resistance and an increase in mean fetal arterial pressure. As arterial pressure is increased, the arterial baroreceptors are stimulated, heart rate is reflexly slowed and cardiac output is reduced.

All the components of this response appear to be in place in the mature fetus. Baroreceptors and chemoreceptors are known to be functional in the fetus and to respond to hypoxaemic stimuli (Hanson 1988). When α-receptors are blocked with pharmacological antagonists the response to hypoxaemia is markedly different. Fetal arterial pressure is not increased and heart rate and cardiac output are augmented rather than being depressed (Reuss et al 1982, Jones &

Ritchie 1983). Yet the degree to which this 'adrenergic mechanism' is responsible for fetal responses to hypoxic stress is unknown. There is no doubt that a redistribution of cardiac output occurs during hypoxaemia even with autonomic nervous system blockade. This suggests that local autoregulatory mechanisms are powerful determinants of flow during hypoxaemic periods, especially for those organs whose flow is altered in the presence of adrenergic receptor blockade, though other circulating factors may also be important. Examples of substances whose concentrations increase during hypoxaemia include: catecholamines, arginine vasopressin, adrenocorticotropin (ACTH), gluco-corticoids, plasma renin activity, β-endorphin and atrial natriuretic peptide (see Rudolph 1984). Nevertheless, the role of these substances as part of the fetal arsenal designed to combat the ill effects of reduced oxygen supply is not clear.

Peeters et al (1979) have demonstrated a complex inverse relationship between organ blood flow and oxygen content for fetal heart and brain, whereas a positive relationship is found for lung. The autoregulatory mechanism appears to be dependent on many factors other than oxygen; hydrogen ion concentration and CO_2 tension are likely to be important flow regulators also. However, this local mechanism appears to be present in the sheep fetus in mid gestation (Iwamoto et al 1989). It is therefore perhaps not accidental that the organs that show the greatest autoregulatory response are those spared during hypoxaemic episodes and those least affected in asymmetrical IUGR (Warshaw 1986). The physiological basis for this powerful and important mechanism is unknown and requires intensive investigative effort.

Cardiac responses to hypoxaemia

We have investigated the effect of moderate hypoxaemia on fetal cardiac function during acute hypoxaemia (Reller et al 1989) and during 'spontaneous' chronic hypoxaemia (Reller et al 1986). Cardiac function was not altered during moderate chronic hypoxaemia ($P_{O2} < 17$ mmHg) but in acute studies of fetal sheep made hypoxaemic, right but not left ventricular stroke volume was significantly reduced when hypertension was present under conditions of β-adrenergic and cholinergic blockade (Reller et al 1989). Furthermore, when the hypertension was removed by administering nitroprusside, right ventricular stroke volume returned to normal. In normoxaemic fetuses, increases in arterial pressures depress the stroke volumes of both ventricles (Thornburg et al 1987), though the right ventricle is more easily depressed by increased arterial pressure than is the left (Reller et al 1987). These experiments suggest that at least part of the reduction in stroke volume seen in acute hypoxaemia is related to increased myocardial afterload and that under such circumstances the right ventricle, because of its unique sensitivity, bears most of the reduction in stroke volume (Fig. 5).

Placental responses to stress

Because the placenta is part of the fetal organism, placental responses to intrauterine stress should be considered as part of any pathophysiological process causing stress to the fetus. Unfortunately, the topic is too extensive to cover here, apart from a few comments. First, placental blood flow is generally maintained during acute oxygen deprivation stresses. This maintenance of flow is undoubtedly a crucial survival mechanism for the fetus. However, abnormal umbilical flows or pulsatility indices are known to be associated with abnormal blood gas levels (Ferrazzi et al 1988) and growth retardation (Burke et al 1990). Furthermore, abnormal velocity waveforms are related to placental vascular abnormalities in experimental animals (Morrow et al 1989). Second, altered placental structure is associated with retarded fetal growth. For example, morphometric analyses indicate that the placentas of small-for-gestational-age babies are proportionately small with a significantly smaller functional surface area (Teasdale 1984). Third, placental function is affected in babies with abnormal growth. For example, depressed placental transport of amino acids in growth-retarded humans (Cetin et al 1988, Dicke & Henderson 1988) and depressed calcium ion transport in growth-retarded rats (Mughal et al 1989) have been demonstrated. This suggests that the placenta may participate in the pathological process, even though it does not point to the origin of the defect. Lastly, because the placenta is a highly metabolically active organ, it may give priority to its own needs for oxygen and nutrients, above those of the fetus (Owens et al 1989).

Summary

The data presented here indicate that the fetus has powerful adaptive responses to oxygen deprivation stress, using baroreceptor and chemoreceptor reflexes as well as adrenally derived circulating catecholamines and autoregulatory mechanisms. A better understanding of the fetal responses to chronic stresses is needed. This knowledge is crucial to our understanding of the relationship between the first line survival defences described here, and the differential growth of the fetal organism with its life-long sequelae.

Acknowledgements

The author thanks Drs Mark Morton, Mark Reller, C. Wright Pinson, Deborah Reid and George Giraud, who conducted the many fetal experiments from our laboratory mentioned here. Technical support was given by Tom Green, Bob Webber, Pat Renwick, Paul Klas, Dana McNaught and Sharon Norris. Dr Reller was supported by a Clinical Investigator Award from the National Institutes of Health. Supported by NIH grants HD10034-13 and HL43015 and the American Heart Association/Oregon.

References

Anderson DF, Bissonnette JM, Faber JJ, Thornburg KL 1981 Central shunt flows and pressures in the mature fetal lamb. Am J Physiol 241:H60–H66

Ashwal S, Majcher JS, Nestor V, Longo L 1980 Patterns of fetal lamb regional cerebral blood flow during and after prolonged hypoxia. Pediatr Res 14: 1104–1110

Bocking AD, Gagnon R, White SE, Homan J, Milne KM, Richardson BS 1988 Circulatory responses to prolonged hypoxemia in fetal sheep. Am J Obstet Gynecol 159:1418–1424

Burke G, Stuart B, Crowley P, Ní Scanaill S, Drumm J 1990 Is intrauterine growth retardation with normal umbilical artery blood flow a benign condition? Br Med J 300:1044–1045

Cetin I, Marconi AM, Bozzetti P et al 1988 Umbilical amino acid concentrations in appropriate and small for gestational age infants: a biochemical difference present in utero. Am J Obstet Gynecol 158:120–126

Clark KE, Skillman CA 1984 Methods of acute and chronic reduction of uteroplacental blood flow. In: Nathanielsz PW (ed) Animal models in fetal medicine, no. III. Perinatology Press, Ithaca, NY, p 37–57

Cohn HE, Sacks EJ, Heymann MA, Rudolph AM 1974 Cardiovascular responses to hypoxemia and acidemia in fetal lambs. Am J Obstet Gynecol 120:817–824

Dawes GS 1968 Foetal and neonatal physiology. Year Book Medical Publishers, Chicago

Dicke JM, Henderson GI 1988 Rapid communication: placental amino acid uptake in normal and complicated pregnancies. Am J Med Sci 295:223–227

Ferrazzi E, Pardi G, Bauscaglia M et al 1988 The correlation of biochemical monitoring versus umbilical flow velocity measurements of the human fetus. Am J Obstet Gynecol 159:1081–1087

Hanson MA 1988 The importance of baro- and chemoreflexes in the control of the fetal cardiovascular system. J Dev Physiol 10:491–511

Itskovitz J, LaGamma F, Rudolph AM 1982 Effects of hemorrhage on umbilical venous return and oxygen delivery in fetal lambs. Am J Physiol 242:H543–H548

Itskovitz J, LaGamma F, Rudolph AM 1987 Effects of cord compression on fetal blood flow distribution and O_2 delivery. Am J Physiol 252:H100–H109

Iwamoto HS, Rudolph AM, Mirkin BL, Keil LC 1983 Circulatory and humoral responses of sympathectomized fetal sheep to hypoxemia. Heart Circ Physiol 14:H767–H772

Iwamoto HS, Kaufman T, Keil LC, Rudolph AM 1989 Responses to acute hypoxemia in fetal sheep at 0.6–0.7 gestation. Am J Physiol 256:H163–H620

Jones CT, Ritchie JWK 1983 The effects of adrenergic blockade on fetal response to hypoxia. J Dev Physiol (Oxf) 5:211–222

Longo LD, Wyatt JF, Hewitt CW, Gilbert RD 1976 A comparison of circulatory responses to hypoxic hypoxia and carbon monoxide hypoxia in fetal blood flow and oxygenation. In: Longo LD, Reneau DD (eds) Fetal and newborn cardiovascular physiology. Garland STPM Press, New York, p 259–288

Morrow RJ, Adamson SL, Bull SB, Ritchie JWK 1989 Effect of placental embolization on the umbilical arterial velocity waveform in fetal sheep. Am J Obstet Gynecol 161:1055–1060

Mughal MZ, Ross R, Tsang R 1989 Clearance of calcium across *in situ* perfused placentas of intrauterine growth-retarded rat fetuses. Pediatr Res 25:420–422

Owens JA, Robinson JS 1988 The effect of experimental manipulation of placental growth and development. In: Cockburn F (ed) Fetal and neonatal growth. Wiley, Chichester, p 49–77

Owens JA, Owens PC, Robinson JS 1989 Experimental growth retardation: metabolic and endocrine aspects. In: Gluckman PD, Johnston BM, Nathanielsz PW (eds) Advances in fetal physiology. Reviews in honor of G. C. Liggins. Perinatology Press, Ithaca, NY, p 263–286

Peeters LLH, Sheldon RE, Jones MD Jr, Makowski EL, Meschia G 1979 Blood flow to fetal organs as a function of arterial oxygen content. Am J Obstet Gynecol 135:637–646

Reller MD, Morton MJ, Thornburg KL 1986 Right ventricular function in the hypoxemic fetal sheep. J Dev Physiol (Oxf) 8:159–166

Reller MD, Morton MJ, Reid DL, Thornburg KL 1987 Fetal lamb ventricles respond differently to filling and arterial pressures and to *in utero* ventilation. Pediatr Res 22:621–626

Reller MD, Morton MJ, Giraud GD, Reid DL, Thornburg KL 1989 The effect of acute hypoxemia on ventricular function during β-adrenergic and cholinergic blockade in the fetal lamb. J Dev Physiol (Oxf) 11:263–269

Reuss ML, Rudolph AM 1980 Distribution and recirculation of umbilical and systemic venous blood flow in fetal lambs during hypoxia. J Dev Physiol (Oxf) 2:71–84

Reuss ML, Parer JT, Harris JL, Krueger TR 1982 Hemodynamic effects of alpha-adrenergic blockade during hypoxia in fetal sheep. Am J Obstet Gynecol 142:410–415

Rudolph AM 1984 The fetal circulation and its response to stress. J Dev Physiol (Oxf) 6:11–19

Rudolph AM, Heymann MA 1970 Circulatory changes during growth in the fetal lamb. Circ Res 26:289–299

Teasdale F 1984 Idiopathic intrauterine growth retardation: histomorphometry of the human placenta. Placenta 5:83–92

Thornburg KL, Morton MJ, Pinson CW, Reller MD, Reid DL 1987 Anatomic and functional distinctions between the fetal heart ventricles. In: Lipshitz J, Maloney J, Nimrod C, Carson G (eds) Perinatal development of the heart and lung. Perinatology Press, Ithaca, NY. (Proc 1st Int Christie Conference) p 49–71

Warshaw JB 1986 Intrauterine growth retardation. Pediatrics in Review 8:PIR107–PIR114

DISCUSSION

Suomi: To what extent do these sets of principles hold at various points during fetal development? Presumably very early in gestation not all of these systems of response are operating; do we know anything about their relative timing?

Thornburg: The studies I described here were all done late in gestation. From the few studies done in sheep in early gestation (Iwamoto et al 1989), we can say that at about 0.6 gestation the autonomic reflexes begin to appear and that hypoxaemic stresses before that time cause *increases* in heart rate rather than decreases. So the conclusions I have suggested here hold only for the late gestation sheep fetus. The very early (first third of pregnancy) sheep fetus has not been studied.

Suomi: The stresses that you described have to do with altering blood flow rates and other internal manipulations. What is known about acute stresses delivered externally to the mother that might increase glucocorticoid output, either indirectly from the mother or by stimulating the adrenal system and the catecholamines directly in the fetus, to produce short-term versions of the same general effect?

Thornburg: Several people have looked at the roles of prostaglandins that might be released within the placental site itself and that would affect the transfer of oxygen by altering placental perfusion. The studies I presented here were chosen because they demonstrate most clearly the redistribution of fetal cardiac output. It is difficult to think of a model that would imitate the acute stresses that might occur where substances (other than catecholamines) would cause vasoconstriction and you would get the same kind of result as you do from clamping the uterine artery. This is a very good question, but we are ignorant about the answer.

Suomi: Do we know anything about the postnatal consequences of the vasoconstriction? Are set points affected, for example?

Thornburg: Drs Deborah Anderson and Job Faber in Portland, Oregon are trying to determine how arterial blood pressure is regulated in the fetus and re-set in early postnatal life. We don't yet know whether sheep fetuses that are subjected to hypoxia have different blood pressures later in life, but this needs to be studied.

Hanson: You concentrated on the acute stress to the fetus, but in terms of placental insufficiency one should be thinking about chronic hypoxaemia. The studies done in the sheep have shown that the redistribution after an acute hypoxic challenge is not maintained if the challenge becomes chronic (see Hanson 1991 for review); the fetal heart rate and blood pressure return to normal, the fetus begins to breathe again, electrocortical activity cycles as normal, and the increase in cerebral blood flow tails off. In the human, cordocentesis suggests that such fetuses are still hypoxic (Soothill et al 1986). The return to control levels is happening even at a time when erythropoietin levels are increased but the haemoglobin level is not, so somehow the fetus is adjusting very rapidly and behaviour is returning to baseline. What is going on?

Thornburg: Two kinds of experiments can be performed to study the chronically hypoxic animal. One type is the sort that I described in my paper, where acute responses to hypoxaemia are determined. But because fetal compensatory mechanisms are so effective, the fetus will generally be able to adapt to some degree, if the experiments are extended in time.

There have been a few experiments of a second type where a stimulus is continuously applied even as the fetus compensates to maintain the status quo; for instance, Kitanaka et al (1989) have put a pregnant ewe in a chamber with low P_{O2} for long periods of time; under these conditions the fetus is able to compensate almost perfectly after a few days, so that the fetus is no longer hypoxaemic. The ewe continues to have a low arterial P_{O2}, but the fetus has a normal value, or almost so. Thus it's difficult to make an animal model which demonstrates the chronic experiments that Nature performs in people. The

Bocking-type experiments come closest because, in them, the uterine artery can be clamped more severely as the experiment proceeds.

Barker: So you both agree that these phenomena are acute changes and revert within days to normal values? Or do they?

Thornburg: Mark and I may disagree slightly on this. I think that the typical fetal responses are demonstrated most clearly in the acute situation. However, many of these adaptations become almost invisible over time. The degree to which they are still operating during long-term episodes of hypoxic stress is not known. My guess is that stress-related redistribution of blood flow is largely mediated by autoregulatory mechanisms and is present throughout any hypoxic period. This offers an explanation for the asymmetrical growth that we see in certain types of growth retardation.

Barker: So whereas the heart rate and blood pressure return to normal, the blood flow continues to be redistributed away from the carcass?

Thornburg: That's what I suspect. Mark Hanson may have experimental data supporting another view. It is a difficult issue.

Hanson: I am sure we both agree there's an urgent need for more work!

Thornburg: Yes, we do!

Lucas: In chronic fetal hypoxaemia there will be an asymmetry that involves reduced growth of the gut. Postnatally, that would seem to be a potential mechanism for reducing nutrient retention. How long does it take in fetal models to get the gut to catch up its growth—in other words, to restore symmetry after asymmetrical growth?

Thornburg: The only study here is Bocking's study in which the uterine artery supplying the fetal sheep was clamped for several days (Bocking et al 1988). In that case the gut blood flow was already back to normal by the end of the experimental period.

Lucas: But what happens if you are actually born asymmetrical—a child with a very tiny gut? That could provide a 'window in time' for suboptimal nutrition.

Thornburg: Catch-up growth during postnatal life, in individuals that have suffered asymmetrical growth retardation, is an area ripe for study. I'm sorry I can't answer your interesting question.

Moxon: I am trying to link the two papers we have heard. The striking finding by Professor Barker of a large placenta in relationship to low birth weight, and perhaps maternal undernutrition, suggests that this might be in some way an adaptive response. In terms of stress and placental function, you, Dr Thornburg, have indicated the extent to which fetal compensatory mechanisms are adapting to preserve blood flow to the placenta, to the point where this might be a drain on the fetus itself. What then are the critical functions of the placenta, to which the hypertrophy is owed? I can see how the enlargement of the placenta could be an encumbrance to the fetus, but perhaps there are placental functions that are as yet unidentified—the secretion of a hormone or a production of a substance—which might fit in with the adaptive response.

Thornburg: This sounds like a matter that needs investigation! We know so little about what controls the growth of the placenta; we know only the general relationship between placental size and fetal size. In any animal model, where uterine artery blood flow is restricted and the fetus becomes smaller, the placenta also becomes smaller. It seems very difficult to make an experimental model where the fetus is small and the placenta is large. We need to design a study where we can make this happen if we want to look at this issue.

Moxon: What known functions of the placenta as an organ might provide a hypothetical basis for this adaptive response with a large placenta?

Thornburg: There are two separate issues. First, a normal placenta has more surface area than is required to provide good oxygenation to the fetus. Second, there appears to be cross-talk between the maternal and fetal circulations, so that the two blood flows are matched properly (Clapp 1989). This means that with a defective placenta, the mechanism that was responsible for flow-matching might be abnormal. Prostaglandins (with other eicosanoids) have been suggested as the most likely candidates for signalling substances (Rankin & McLaughlin 1979), but the full picture isn't clear yet.

Moxon: The idea, in at least a subpopulation of women with large placentas and low birth weight infants, of a role of congenital infection would be worth following up.

Wood: Are there any histopathological studies in relation to the models of hypoxic stress that you describe, Dr Thornburg? I am thinking particularly of the brain, which experiences an increased or protected blood flow. In clinical practice, in the low birth weight preterm baby, some of the circulatory events which may take place, such as bradycardia and hypoxaemia, are not dissimilar from these models, and one is always concerned about periventricular and intraventricular haemorrhages being related to changes in blood pressure and hence cerebral blood flow.

Thornburg: Dr Clapp (1981) found in sheep fetuses that progressive hypoxia and acidosis result in fetal cerebral damage, but that such damage is not necessarily associated with growth retardation.

Smart: Could the relationship that you have suggested between hypoxic stress and what you call asymmetrical growth retardation be applied to the effects of undernutrition during development? Are you suggesting that the same sorts of protective mechanisms operate in developmental undernutrition as in hypoxia?

Thornburg: I wasn't implying that, but it's fair to say that every organ in the body has its own autoregulatory capability, which means that you can take the organ away from the nervous system, perfuse it with blood, and it will regulate its own blood flow, depending on oxygen requirements, etc. This means that there is the possibility, even after birth, that autoregulatory mechanisms might be affected by nutritional state, but I doubt whether anybody has looked at that.

Smart: Presumably it would be relatively easy to investigate?

Thornburg: I agree.

Barker: Could I ask what we know about babies born at high altitude?

Thornburg: They are very small!

Barker: And that's all we know?

Thornburg: Their mothers are relatively hypoxaemic but the fetuses have relatively normal P_{O2} values at birth; the trophoblast is thinner with a smaller surface area than normal, so the gaseous exchange surface area is reduced (M. R. Jackson et al 1987).

Hanson: You are not saying that babies born at altitude show a higher incidence of asymmetrical growth retardation?

Thornburg: No.

Martyn: The responses that you described to hypoxia show two components; one is the diversion of better-oxygenated blood to the brain, which seems a sensible, protective thing for the animal to do. But the other component of the response, the part which you can prevent with α-adrenergic blockage, seems to be rather less clearly beneficial, in that cardiac output and heart rate go down and blood pressure goes up. Do you see that as a protective mechanism, or as something that the fetus would be better off without?

Thornburg: I see it as a protective mechanism from two perspectives. Perhaps hypertension is important as a mechanism to maintain perfusion of the placenta, because placental blood flow is directly proportional to arterial pressure. The fetus protects its arterial pressure to maintain oxygenation. Secondly, when the heart does less work, it requires less oxygen. So, even though heart work is increased as arterial pressure goes up, a slower heart rate keeps myocardial oxygen demands at a minimum under the circumstances. So one can argue that each of the fetal responses has potential protective value.

Murray: Nearly all human neurodevelopmental disorders are commoner in males. Can we expect to find sex differences in fetal responses to hypoxia, or do we have to look elsewhere for the cause of this male excess?

Thornburg: I don't know! But you have asked a very good question.

Barker: Is the fetal heart rate different in the human male and female fetus? I thought that it was lower in males.

Chandra: Blood pressure is higher in boys, even in younger age groups.

Golding: In Professor Barker's population studies, maternal smoking is unlikely to have contributed much to the birth weight and infant growth effects, since few mothers smoked at that time. Nevertheless, the effect of maternal smoking is to produce a symmetrically small fetus (D'Souza et al 1981)—in other words, the insult has been going on over a long time. Mothers who smoke also have relatively large placentas (Wingerd et al 1976) with numbers of cells in these placentas lower than in the normally grown placenta, but the cell sizes are much larger.

Thornburg: The fetuses of smoking mothers are symmetrically small with normal-sized placentas, but the placenta *is* affected. Cytomorphometry has been used to look at placental surface areas at the various levels of villous arborization (van der Velde et al 1983, van der Veen & Fox 1982). These studies show that the number of cytotrophoblast cells is increased and that the diffusion barrier is thickened.

Dobbing: Incidentally, much of the placenta is syncytial, so we should speak of numbers of nuclei rather than numbers of cells.

Golding: I agree!

Barker: In our data, the babies born with large placentas were *asymmetrically* small. Secondly, as Jean Golding has said, the generation of women having babies before the Second World War were not smokers; it was war time that led to a rise in smoking in women. Smoking needs to be borne in mind, but I would be surprised if it was the explanation of our findings.

Thornburg: It may not answer the questions that you pose, but it does bring up a new point about how fetal growth is affected by external mechanisms.

Casaer: Is the redistribution of blood flow always a 'good thing'? In a recent study we could demonstrate that newborns with severe post-asphyxial encephalopathy had relatively high mean flow velocities and low pulsatility indices, both before and after transfusion. But, until one week after the acute insult, these infants were unable to adjust their brain circulation according to changes in haemoglobin concentration; their mean flow velocity values remained high. This long-lasting increased flow thus seems not to be a beneficial adaptation but is very probably a sign of decreased autoregulation (Ramaekers & Casaer 1990).

Thornburg: You may well be right. It's easy to assume that these adaptive responses are good for the fetus. It may be that some of the responses to hypoxic stress, while important for maintaining blood flow acutely, may also be deleterious in the long term. The gut situation mentioned by Dr Lucas may be another example where, if the carcass, kidney and gut are all being sacrificed to spare the brain and heart, there can be long-term deleterious sequelae. There may be survival value, but these are drastic measures for the organs not receiving their full complement of blood flow.

Barker: This is an interesting point, that the fetus may be solving an immediate problem in the best way it can, but the solutions may in the longer term be disastrous.

How is the adrenal sparing mediated, Dr Thornburg?

Thornburg: That is not certain, because the proper experiment is to see whether there is the same redistribution of blood when the fetal sheep has been sympathectomized and when it has not; such an experiment has not been reported. I would guess that it is probably an autoregulatory mechanism, as with the other organs that are spared. Vinson et al (1985) have suggested that adrenal blood flow may be affected by ACTH.

Barker: It seems intuitively sensible to spare the brain and heart but it is less clear why the adrenals should be spared.

Thornburg: It is presumably because the catecholamine response is such an integral part of the hypoxic response that the fetal adrenal gland must maintain its blood supply in order to keep this response intact.

Dobbing: The experiment that Jim Smart (p 32–33) suggests is very seminal to all this, for it seems clear that the relative sparing of the different organs and tissues, in postnatal undernutrition as well as in prenatal, shows a similar pattern to the one described for hypoxia. It has always been a mystery why, in developmental undernutrition, the brain, heart and adrenals should be spared. Nobody appears to have thought of a control on the *supply* of nutrients to the tissues as being responsible for 'sparing'. Most people seem to think in terms of something inherent in the metabolic demands of the tissues, a property of the tissues themselves, which determines it. This could be solved by the experiment of seeing whether, in postnatal undernutrition, there is any parallel in your terms, with oxygen supply, blood flow, and so forth, in those tissues. This is of more than passing interest.

Secondly, I have no trouble in thinking of small fetuses and small placentas going together. Where I have trouble is the assumption that the relationship is causative rather than associative. Is there any good evidence that it's causative? By and large, placental size and fetal size are assumed to go together, since the feto-placental unit may be expected to grow as a unit (Sands & Dobbing 1985).

Thornburg: A relevant experiment is that of Owens & Robinson (1988), who reduced placental mass in sheep by taking placental cotyledons away. This produced smaller fetuses and growth retardation. However, I don't think the available data indicate which way we ought to be thinking here. We need to think of the definitive experiment to do.

Dobbing: It is easy to think in terms of a harmony of the fetus and the placenta as a growth unit, albeit with its subsections with different properties, and 'therefore' the placenta will be the same size. Placental weight is closely related to that of the fetus in nearly all circumstances, not only in growth-retarded fetuses but also in those within the normal range, which is perhaps more important (Sands & Dobbing 1985).

Barker: Dr Thornburg, can you harmonize this view of placenta and fetus as one growth unit with what you said about the ability of the placenta to take priority over the fetus?

Thornburg: It certainly appears to be one unit, but there are exceptions where it doesn't behave entirely that way. It is obvious in Professor Robinson's experiments (late gestation sheep fetuses with reduced placental mass) that the placenta utilizes amino acids at the expense of the fetus. This may be required to maintain the surface area of the remaining cotyledons. It may, teleologically speaking, be a wise move for the placenta to maintain itself in order to keep the fetus alive. In that case there is a difference between fetal and placental growth.

Hanson: Another difference is that the placenta, in the sheep anyway, does not grow at the same rate as the fetus, and placental growth early in gestation somehow dictates fetal growth rates later on.

Mott: Dr Thornburg, you found that catecholamines were maintaining an essential component of the fetal response to hypoxia. What about cortisol, as a longer-term type of hormonal response?

Thornburg: Cortisol levels are dramatically increased, some 5–10-fold, during hypoxaemic episodes (Challis et al 1986), but they decrease with time up to 6–7 hours. Long-term responses are unknown.

Suomi: You showed that in the sheep the relative sparing of adrenals, heart and brain maintained a certain proportion across the different types of hypoxic condition tested. Would you expect those sorts of relative proportions to hold for primates, or, for example, might you expect to see even greater sparing of brain areas there?

Thornburg: The response to hypoxaemia has been studied in rhesus monkeys by B.T. Jackson et al (1987). They saw similar responses to hypoxaemia to those found in sheep.

Suomi: Could part of the priority given to the placenta over the fetus be a matter of timing? That is, could it be that if you looked relatively early under stress conditions you might see this sparing of the adrenals, heart and brain, but over a longer period of time you could start to see things getting out of synchrony? It might depend at what point in the process you are collecting your data.

Thornburg: In the study by Owens & Robinson (1988) there was no priority for the placenta until fairly late in gestation, when the fetus was getting rather large and appeared to be 'running out' of placental surface area.

Suomi: Is the effect also perhaps independent of gestation—that is, simply in terms of the duration of the stress itself?

Thornburg: Perhaps so.

References

Bocking AD, Gagnon R, White SE, Homan J, Milne KM, Richardson BS 1988 Circulatory responses to prolonged hypoxemia in fetal sheep. Am J Obstet Gynecol 159:1418–1424

Challis JRG, Richardson BS, Rurack D, Wlodeck ME, Patrick JE 1986 Plasma adrenocorticotropic hormone and cortisol and adrenal blood flow during sustained hypoxemia in fetal sheep. Am J Obstet Gynecol 155:1332–1336

Clapp JF III 1981 Neuropathology in the chronic fetal lamb preparation: structure–function correlates under different environmental conditions. Am J Obstet Gynecol 141:973–986

Clapp JF III 1989 Utero-placental blood flow and fetal growth. In: Sharp F, Fraser RB, Milner RDG (eds) Fetal growth. Peacock Press, Ashton-Under-Lyme (Proceedings 20th Group of the Royal College of Obstetricians and Gynaecologists)

D'Souza SW, Black P, Richards B 1981 Smoking in pregnancy: associations with skinfold thickness, maternal weight gain, and fetal size at birth. Br Med J 282:1661–1663

Hanson MA 1991 The control of the fetal circulation. In: Hanson MA (ed) The fetal and neonatal brain stem—developmental and clinical issues. Cambridge University Press, Cambridge, in press

Iwamoto HS, Kaufman T, Keil LC, Rudolph AM 1989 Responses to acute hypoxemia in fetal sheep at 0.6–0.7 gestation. Am J Physiol 256:H613–H620

Jackson BT, Piasecki GJ, Novy MJ 1987 Fetal response to altered maternal oxygenation in rhesus monkey. Am J Physiol 252:R94–R101

Jackson MR, Mayhew TM, Haas JD 1987 Morphometric studies on villi in human term placentae and the effects of altitude, ethnic grouping and sex of newborn. Placenta 8:487–495

Kitanaka T, Alonso JG, Gilbert RD, Sui BL, Clemons GK, Longo LD 1989 Fetal responses to long-term hypoxemia in sheep. Am J Physiol 256:R1348–R1354

Owens JA, Robinson JS 1988 The effect of experimental manipulation of placental growth and development. In: Cockburn F (ed) Fetal and neonatal growth. Wiley, Chichester, p 49–77

Ramaekers VT, Casaer P 1990 Defective regulation of cerebral oxygen transport after severe birth asphyxia. Dev Med Child Neurol 32:56–62

Rankin JHG, McLaughlin MK 1979 The regulation of the placental blood flows. J Dev Physiol 1:3–30

Sands J, Dobbing J 1985 Continuing growth and development of the third-trimester human placenta. Placenta 6:13–22

Soothill PW, Nicolaides KH, Bilardo CM, Campbell S 1986 Relation of fetal hypoxia in growth retardation to mean blood velocity in the fetal aorta. Lancet 2:1118–1120

van der Veen F, Fox H 1982 The effects of cigarette smoking on the human placenta: a light and electron microscopic study. Placenta 3:243–256

van der Velde WJ, Copius Peereboom-Stegeman JHJ, Treffers PE, James J 1983 Structural changes in the placenta of smoking mothers: a quantitative study. Placenta 4:231–240

Vinson GP, Pudney JA, Whitehouse BJ 1985 The mammalian adrenal circulation and the relationship between adrenal blood flow and steroidogenesis. J Endocrinol 105:285–291

Wingerd J, Christianson R, Lovitt WV, Schoen EJ 1976 Placental ratio in white and black women: relation to smoking and anemia. Am J Obstet Gynecol 124:671–675

Programming by early nutrition in man

Alan Lucas

MRC Dunn Nutrition Unit, Downhams Lane, Milton Road, Cambridge CB4 1XJ, UK

Abstract. Whether early diet influences long-term health or achievement is a key question in nutrition. Such long-term consequences would invoke the concept of 'programming'—a more general process whereby a stimulus or insult at a critical period of development has lasting or lifelong significance. Data from small mammals and primates show that early nutrition may have potentially important long-term effects, for example on blood lipids, plasma insulin, obesity, atherosclerosis, behaviour and learning. Corresponding studies in man have been largely retrospective and difficult to interpret. The preterm infant is however an important model for human research because formal random assignment to early diet is practical. A large prospective randomized multicentre study has been undertaken on 926 preterm infants to test the hypothesis that early diet influences long-term outcome. Diets included human milk, standard formula and nutrient-enriched preterm formula. The diet consumed for on average the first month *post partum* had a major impact on subsequent developmental attainment, growth and allergic status in early childhood. That such a brief period of dietary manipulation has lasting significance implies that the neonatal period is critical for nutrition after preterm birth. These data may have broader implications for human nutrition.

1991 The childhood environment and adult disease. Wiley, Chichester (Ciba Foundation Symposium 156) p 38–55

Despite intensive research on infant nutrition throughout this century, uncertainty persists over nearly every major area of practice. A key factor in this uncertainty has been a lack of knowledge on whether diet or nutritional status in early life has a long-term or permanent influence on health, function or achievement. Until clinical scientific information of this nature is established, the most appropriate nutritional management of sick children in hospital and the most prudent advice for healthy infant feeding practice in the community will remain speculative and prey to changing fashion.

The concept that nutrition in infancy could have long-term significance, however, raises questions of fundamental biological importance. Is it plausible or, arguing teleologically, evolutionarily likely, that such a brief period of life could be a critical one for nutrition? How strong is the evidence that other early metabolic events have long-term consequences? What is the nature of the triggering mechanisms for these events? If the consequences of an early stimulus or insult might not be expressed until later in life, how and where is the 'memory' of it stored in the meantime, while the cells of the body are constantly replicating or being replaced? And is there indeed hard evidence that early nutrition influences long-term outcome in man? The purpose of this brief review is to give consideration to these important issues.

The concept of early biological programming

There are a number of ways in which an early event (nutritional or otherwise) could have permanent or long-term consequences. Three general categories are proposed, as follows:

1. Direct damage (for example, loss of limb due to vascular accident or trauma).
2. Induction, deletion, or impaired development of a permanent somatic structure as the result of a stimulus or insult operating at a critical period.
3. Physiological 'setting' by an early stimulus or insult at a 'sensitive' period, resulting in long-term consequences for function; the effects could be immediate or deferred.

In this article I shall refer to the second and third of these three processes as 'programming', though I accept that others might use the term in a more limited sense. As a working definition, then, 'programming' occurs when an early stimulus or insult, operating at a critical or sensitive period, results in a permanent or long-term change in the structure or function of the organism. An essential component of this concept is the notion of a 'sensitive' or 'critical' period, and it is for this reason that direct damage, as indicated in category 1 above, has been excluded. (Even so, the traumatic loss of a limb, say, at a critical period of development could itself have a programming effect on subsequent behaviour and cerebral function.) The proposed definition of programming does include damage to a structure or function when the insult (for instance, inadequate nutrition) would exert its effects only if applied during a sensitive period.

Evidence for programming in general

Outside the field of nutrition, evidence for programming is legion. It is not the purpose of this article to review these extensive data, but a few key examples are cited.

Programming may occur as a normal part of biological development or in response to unphysiological events. It may occur under the influence of genetically determined internal triggers, or as a result of exogenous stimuli. For example, at a critical period in male fetal rats, the brain is programmed for male sexual behaviour by the endogenous release of testosterone from the developing testis. At this stage, a single exogenous dose of testosterone administered experimentally to a female fetal rat will permanently reorientate sexual behaviour (Angelbeck & DuBrul 1983).

Hormones indeed have a well-established role as programming agents. Mullerian inhibitory factor, secreted at a critical period by the fetal testis, induces regression of the Mullerian ducts and thus permanent deletion of female internal genitalia (Jost 1972). In contrast, testosterone secreted locally from the testis induces the development of male internal genitalia from the Wolffian duct on the same side of the body—thus, unilateral anorchia would deprive the Wolffian system on that side of an essential programming signal. Thyroxine is critical for early brain development with transient deficiency resulting in permanent changes that would not be seen later in life (Brasel & Boyd 1975). Conversely, excess of thyroxine in the neonatal period in experimental animals resets the pituitary–hypothalamic axis with respect to thyroid-stimulating hormone responses in later life (Besa & Pascual-Leone 1984). The substantial evidence that a wide variety of hormones and growth factors operate at critical periods raises the possibility, discussed below, that an intermediary role of hormones could explain some long-term effects of early nutrition.

Other agents which have a known programming influence include sensory stimuli (for example, in the development of the visual pathway), antigens (in immunological development) and drugs. A single dose of phenobarbitone given to a newborn rat will programme a life-long change in cytochrome P450-dependent monooxygenase activity (Bagley & Hayes 1983), highlighting the exquisite sensitivity that may exist to early programming stimuli.

Nutritional programming in animals

Animal studies have shown that nutrition may programme long-term outcome. Hahn (1984) manipulated litter size in rats to produce litters of four or 14 pups. In small litters the pups were overfed during the short suckling period. In adulthood the previously overfed rats developed raised plasma cholesterol and insulin concentrations. Rats weaned onto a high carbohydrate diet also showed these changes, and, in addition, the early dietary manipulation produced, in adulthood, increased activities of two key enzymes in lipid biosynthetic pathways: fatty acid synthetase and HMG CoA reductase (important in fat and cholesterol synthesis).

Mott and Lewis (see elsewhere in this symposium: Mott et al 1991) have shown in primates that early overfeeding may programme later obesity and that breast-fed and bottle-fed baboons have long-term differences in their lipid metabolism and in the degree of atherosclerosis. These results demonstrate that early diet could play a role in the later development of disease states that have considerable significance in humans, and they emphasize the need for outcome studies in man. The primate studies on early overfeeding and later obesity also indicate that dietary programming effects are not necessarily immediate, but may emerge later in life, suggesting that human studies in this field must be prolonged.

In addition to the work on early diet and later metabolism and morbidity, a large body of animal data, principally from studies on rats, has indicated that the quality of nutrition at a 'vulnerable' period of early brain development could have permanent consequences for brain size, brain cell number and performance (Dobbing 1981). Katz (1980) and Smart (1977) have reviewed the extensive literature on whether early undernutrition influences later learning and memory. In Smart's (1986) review of 165 animal experiments, the number of studies in which undernourished animals fared significantly worse than controls ($n = 80$) greatly and significantly outweighed studies coming to the opposite conclusion ($n = 12$). Although there may be a reporting bias for positive studies, it is likely that these data reflect long-term consequences of early underfeeding. The relevance of these studies to man require further work.

Collectively, these and many other animal studies provide convincing evidence that nutrition at a critical or sensitive period in early life may influence a wide variety of metabolic, developmental and pathological processes in adulthood.

Dietary programming in man: previous studies

Unfortunately, the great majority of investigations on the consequences of early nutrition in man have been retrospective and flawed by problems with study design.

The largest category of study in this field relates to early protein-calorie malnutrition and later achievement (Grantham-McGregor 1987). Given the enormous investment made, it is disappointing that firm conclusions cannot yet be drawn. These largely retrospective studies are seriously confounded by the poverty, poor social circumstances and lack of stimulation that accompany malnutrition. Attempts to generate suitable controls have never been satisfactory. Prospective studies are few and randomization to groups is rare. The outcome disadvantages for poorly nourished children have often been small, though both type I and type II statistical errors can be expected, given the difficulties encountered.

The effect of individual nutrients on later brain development is receiving increasing attention. Iron deficiency in infancy, common both in the West and in developing countries, has been shown to relate to poor developmental performance (Ankett et al 1986). Some evidence suggests that subsequent iron supplementation may not prevent later reduction in cognitive ability at five years and that a brief period of relatively mild deficiency could have long-lasting consequences for behaviour and school performance (Walter et al 1989, Dallman et al 1980). Irreversible long-term consequences of early iron deficiency have also been demonstrated in rats (Dallman et al 1975, Yehuda & Youdim 1988).

The possibility that inadequate supplies of long-chain $\omega - 3$ fatty acids in the diets of formula-fed and preterm infants might impair cerebral and retinal development has become an area of concern. The developing brain (notably the cerebral cortex) and retina accumulate large quantities of docosahexaenoic acid ($22:6\omega - 3$) (Carey 1982, Mesami & Timcras 1982). Such long-chain lipids are not present in significant amounts in many formulas, which frequently have low contents of the precursor, linolenic acid ($18:3\omega - 3$). Compelling data in primates (Neuringer et al 1986) now show that an insufficiency of these fatty acids at a critical stage of retinal development results in long-term, irreversible impairment of retinal function. Unpublished studies in premature babies now suggest a similar phenomenon.

A number of investigations have focused on early diet in relation to diseases found in affluent countries. Whether early lipid intake in man influences later lipid status or atherosclerosis has not been established (Lloyd 1987). Early salt intake has not been convincingly related to later blood pressure, despite earlier concerns (Whitten & Steward 1980, Dahl & Love 1957). The long-term consequences of early excessive food intake is an important issue needing further study. Many groups have examined the relationship between infant obesity and later obesity, which in general are weakly correlated (Poskitt & Cole 1977). However, this may not be the best approach. We have used the doubly labelled water method to explore energy metabolism in free-living infants (Lucas et al 1987). Our unpublished data indicate that additional energy in the diet may be expended and not necessarily deposited in body stores. If so, it is early energy intake and not just fatness that needs to be correlated to later fatness and morbidity. Our preliminary investigations show that energy intake at three months of age does not correlate with skinfold thickness at that age, but is significantly correlated with skinfold thickness at two years. These findings are consistent with those of Kramer (1981), who found that, compared with bottle-fed babies, those who were breast fed in infancy received some protection against obesity in later childhood and adolescence, even after adjusting for confounding factors. It is likely that bottle-fed infants at that time consumed more energy than breast-fed babies. Conversely, mothers who were starving in the third trimester during the Dutch Hunger Winter (1944/1945) had children who grew up to be of normal height and intelligence, but were thinner than their peers.

The preterm infant as a model for programming studies

The lack of prospective, long-term outcome studies on diet in man has related to the unattractiveness of this type of work to many investigators, the unwillingness of funding bodies to support it, disbelief that early influences could have lasting effects, and the inherent difficulties in mounting formal randomized longitudinal studies. With regard to the latter point, we have identified a circumstance in man where it is practical and ethical to assign infants randomly to diet and to follow them up long term—that is, after preterm birth.

Between 1982 and 1985 we assigned nearly 1000 preterm infants to diet and studied them intensively in the newborn period. We are now following them up indefinitely (Lucas et al 1984). The babies were randomized at birth, in four parallel trials, to either a preterm formula (enriched in protein, energy, macrominerals and trace nutrients to meet the calculated increased requirements of preterm infants) versus banked donor breast milk, or to preterm formula versus a standard formula. For each comparison these feeds were used as sole diets or as supplements to the mother's own milk. The infants remained on the assigned diets for an average of one month; after that there was no influence on dietary management. Follow-up data are available to 18 months corrected age; a 7–8-year follow-up is in progress.

While the trials are providing essential outcome data that will guide the clinical management of preterm infants, the study also offers a unique opportunity to test, in a human model, whether early diet could influence long-term outcome in man. The principal medium-term outcome response chosen, for calculation of the trial size, was neurodevelopment, but a number of other key outcomes have been explored. In this paper, three examples of outcome data have been selected that support the concept of programming.

Early diet and later allergic reactions

The effect of early diet on later allergy has been much debated. A major problem in the interpretation of results from many studies arises from the lack of random assignment to diet. Clearly, random assignment to breast feeding or formula feeding would be unethical in healthy infants, and yet social and demographic differences found between these two groups confound comparative analyses. In preterm infants, random allocation to human milk or formula is ethical and feasible. In one limb of our trial we compared infants randomly assigned to banked donor breast milk as sole diet or as a supplement to mother's milk (that is, all received only human milk) with those fed a preterm formula as sole diet or supplement (that is, all were exposed to cows' milk formula). Beyond one month, on average, the trial diets were discontinued and there was no significant difference in dietary management between groups. At 18 months the pattern of response depended on whether or not the child had a family history of allergy (Lucas et al 1990a). In infants with no such history, interestingly, those fed previously on cows' milk formula had a small (non-significant) reduction in the incidence of reactions to cows' milk, other food or drugs, and in eczema and wheezing. In contrast, in the smaller subgroup with a positive family history of allergy, babies given cows' milk formula rather than breast milk in the neonatal period had a dramatic increase in these allergic responses, notably in eczema and reaction to food or drugs. Our results (Table 1) demonstrate that

TABLE 1   Family history of atopy, neonatal diet and allergic reactions at 18 months in infants born preterm

Allergic reaction	Family history of atopy			No family history			†Interaction between family history and diet (P value)
	Human milk (n = 38)	Preterm formula (n = 37)	Odds ratio (95% CI)	Human milk (n = 189)	Preterm formula (n = 182)	Odds ratio (95% CI)	
Eczema	16% (6)	41% (15)	3.6* (1.2, 11)	21% (40)	16% (29)	0.7 (0.4, 1.2)	<0.01
Reactions to:							
Cows' milk	3% (1)	5% (2)	2.1 (0.2, 25)	5% (9)	3% (5)	0.6 (0.2, 1.7)	
All foods	11% (4)	22% (8)	2.3 (0.6, 8.3)	10% (18)	9% (16)	0.9 (0.5, 1.9)	
Drugs	3% (1)	16% (6)	7.1 (0.8, 50)	7% (14)	5% (10)	0.7 (0.3, 1.7)	
Wheezing or asthma	21% (8)	30% (11)	1.6 (0.6, 1.4)	24% (40)	22% (40)	0.9 (0.6, 1.5)	
Any of above	34% (13)	65% (24)	3.6* (1.4, 9.1)	46% (86)	42% (76)	0.9 (0.6, 1.3)	<0.001

Infants were divided into two groups according to whether or not there was a family history of atopy. Within each group the percentage incidence of allergic reactions (no. of subjects) is compared in infants randomly assigned to donor milk or preterm formula as sole diets or supplements to mother's milk (thus the donor milk-fed group received only human milk). The odds ratio (95% confidence interval, CI) is recorded for the incidence of allergy on preterm formula versus human milk. Corresponding odds ratios in the two family history groups are also compared, to test for an interaction between family history and diet.
*$P < 0.05$. †Significant interaction indicates that the effect of diet with no atopic family history was significantly different from the dietary effect when the family history was positive; i.e., there was a significant difference in odds ratios between family history groups.

in genetically susceptible individuals, a brief period of dietary manipulation 'programmes' the infant's propensity for developing a wide range of allergic or atopic manifestations.

Early diet and later linear growth

Using multiple regression analysis on our cohort of infants (Lucas et al 1989a) we identified, out of a large number of clinical and demographic factors analysed (over 50), a total of five perinatal factors that were significantly and independently related to body length at 18 months (see Fig. 1). One, of course, was the infant's sex. In addition, being a twin or triplet was associated with a 1 cm reduction in length at 18 months; infants fed exclusively on human milk were also on average 1 cm shorter; being born small for gestation was associated with a 1.3 cm reduction in length, and those having biochemical evidence of severe metabolic

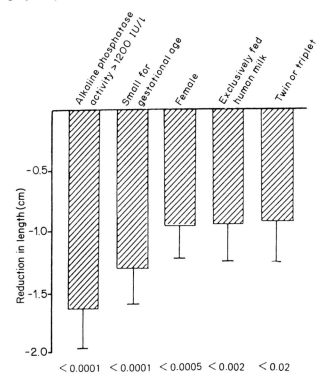

FIG. 1.   Factors independently associated with reduced body length at 18 months after term. The magnitudes (regression coefficients) of the associations and standard errors depicted were derived from a regression model constructed for each age group, with body length as the dependent variable.

bone disease in the newborn period (plasma alkaline phosphatase level over 1200 IU/l) had a major, 1.6 cm reduction. The latter condition, most common in preterm infants fed unsupplemented breast milk, reflects a phosphorus and calcium intake inadequate to meet the rapid bone mineralization that would have occurred *in utero* in the third trimester. The additive effect of the adverse factors for body length described would be to displace the infant's body length from the 50th centile to the 3rd.

It is pertinent that, of all the factors analysed, those few that were related to reduced long-term linear growth were all factors associated with a period of reduced linear growth in the perinatal period as a result of impaired fetal or neonatal nutrition—in the case of bone disease, due specifically to impaired mineral supply.

It seems reasonable to suggest that early nutrition has a programming effect on long-term linear growth, and our preliminary studies made later in childhood continue to support that view. I would speculate, teleologically, that it makes

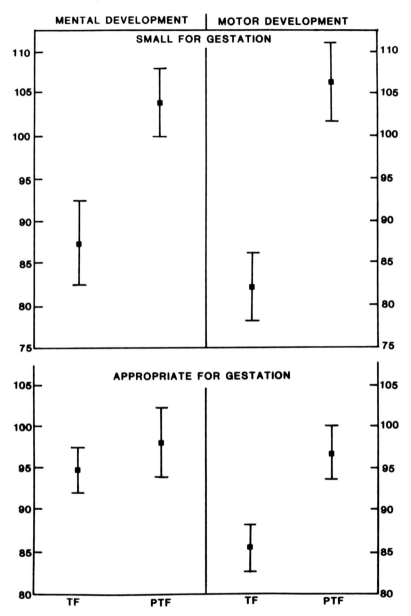

FIG. 2.   Mental and motor development indices (Bayley Scales) at 18 months in infants randomly assigned to a standard 'term' formula (TF) or a special nutrient-enriched 'preterm' formula (PTF) during the early weeks *post partum*, in babies born small (*above*) and appropriately sized for gestation (*below*). Bars represent mean (SE) neurodevelopmental scores. (Adapted from Lucas et al 1990b.)

good sense, from an evolutionary point of view, for an infant to monitor its environment after birth and to set growth projections according to nutrient availability. In an environment with poor nutrient supply, for instance, it would be advantageous to be small. Babies with placental insufficiency or impaired postnatal nutrition due to prematurity might 'interpret' their environment as 'nutrient-poor' and, perhaps irreversibly, set themselves up for shorter stature.

Early diet and later development

Our preterm infant feeding trial has provided the first opportunity to study the effect of diet on later brain development in a large, strictly randomized prospective trial in man. Lucas et al (1989b, 1990b) show that a brief period of dietary manipulation, for on average one month, has a significant impact on developmental quotient in infancy. The effect is most dramatic in small-for-gestational-age infants, who are born already deprived of nutrient supply. Studies at the 18 months follow-up show that infants randomly assigned to a standard term formula rather than a nutrient-enriched 'preterm' formula, just for the early weeks, subsequently had a major deficit in developmental scores (Fig. 2) (Lucas et al 1990b). This was particularly marked for motor development, where the deficit was of the order of one standard deviation, and greater still in infants born small for gestation.

Whether these long-term dietary effects on development relate to the early differences in brain growth observed between diet groups (Lucas et al 1984), or to the lack of a specific critical nutrient or signal for cerebral development, now requires intensive investigation.

Investigation of programming mechanisms

Collectively, the animal studies and the new human data now emerging suggest that the newborn period is a critical one for nutrition. If so, it becomes important to define the mechanisms involved, since these may prove central to an understanding of a major route to adult disease. In this final section, I shall briefly consider key areas for exploration.

Some programming effects are likely to be immediate. An acute insult or lack of a critical signal at the appropriate stage could irreversibly interrupt or change a developmental process. There has been an assumption that early undernutrition might exert its effects simply by failing to fuel a growth process, for example neuronal growth at a critical stage, with subsequent failure of catch-up. Although this might prove to be true, the concept needs careful investigation. It may be that damage is done by the absence of a single critical factor, or even indirectly by a failure to operate a coupling mechanism. As we explore these mechanisms at a cellular level it may prove useful to think of nutrients not as foods, but rather as potential biological triggering agents, and to seek out possible receptors and coupling systems.

Other programming effects, as indicated in some animal studies, are deferred. In this instance the question is how the memory of the early event is stored and later expressed. Three cellular mechanisms might be envisaged. Firstly, the nutrient environment of a developing cell line may permanently alter gene expression, so that cellular progeny will have the 'message' passed on to them. Alternatively, as in the immune system, early nutrition might affect clonal selection. Thus the animal data showing long-term effects of diet on nutrient absorption could be explained if the early pattern of nutrients in the gut lumen influenced which enterocyte lines were selected. Finally, as has been hypothesized for the programming of adiposity, early nutrition could permanently influence cell number. These mechanisms need to be explored, initially in animal models.

As suggested, it is possible that nutrients do not always act directly, but through an intermediary mechanism. Previously, I have speculated that gut hormones are a potential set of intermediary agents (Lucas et al 1985). Their pattern of release is highly related to the pattern of nutrients consumed (for instance, we have reported major differences in gut hormone release between breast-fed and formula-fed infants: Lucas et al 1980); gut peptide receptors are widely distributed throughout the body; we have shown that gut peptide release is dramatic in the neonatal period, which perhaps implies that they have a unique role in early life; and, finally, hormones are known to be potent programming agents in other contexts.

Nutritional programming is a major unexplored area of human biology with broader implications for understanding the impact of early experience on adult disease and achievement.

References

Angelbeck JH, DuBrul EF 1983 The effect of neonatal testosterone on specific male and female patterns of phosphorylated cytosolic proteins in the rat preoptic-hypothalamus, cortex and amygdala. Brain Res 264:277–283

Ankett MA, Parks YA, Scott PH, Wharton BA 1986 Treatment with iron increases weight gain and psychomotor development. Arch Dis Child 61:849–857

Bagley DM, Hayes JR 1983 Neonatal phenobarbital administration results in increased cytochrome P450-dependent monooxygenase activity in adult male and female rats. Biochem Biophys Res Commun 114:1132–1137

Besa ME, Pascual-Leone AM 1984 Effect of neonatal hyperthyroidism upon the regulation of TSH secretion in rats. Acta Endocrinol 105:31–39

Brasel JO, Boyd D 1975 Influence of thyroid hormone on fetal brain growth and development. In: Fisher DA, Burrows GN (eds) Perinatal thyroid physiology and disease. Raven Press, New York

Carey EM 1982 The biochemistry of fetal brain development and myelinisation. In: Jones CT (ed) The biochemical development of the fetus and neonate. Elsevier Biomedical Press, New York, p 287–336

Dahl LK, Love RA 1957 Etiological role of sodium chloride intake in essential hypertension in humans. J Am Med Assoc (JAMA) 164:397–402

Dallman PR, Siimes MA, Manies EC 1975 Brain iron: persistent deficiency following short-term iron deprivation in the young rat. Br J Haematol 31:209–215

Dallman PR, Siimes MA, Stekel A 1980 Iron deficiency in infancy and childhood. Am J Clin Nutr 33:86–118

Dobbing J 1981 Nutritional growth restriction and the nervous system. In: Davison AN, Thompson RHS (eds) The molecular bases of neuropathology. Edward Arnold, London p 221–233

Grantham-McGregor S 1987 Field studies in early nutrition and later achievement. In: Dobbing J (ed) Early nutrition and later achievement. Academic Press, London p 128–174

Hahn P 1984 Effect of litter size on plasma cholesterol and insulin and some liver and adipose tissue enzymes in adult rodents. J Nutr 114:1231–1234

Jost A 1972 A new look at the mechanisms controlling sex differentiation in mammals. Johns Hopkins Med J 130:38–53

Katz HB 1980 The influence of undernutrition on learning performance in rodents. Nutr Abstr Rev 50:767–783

Kramer MS 1981 Do breast-feeding and delayed introduction of solid food protect against subsequent obesity? J Pediatr 98:883–887

Lloyd JK 1987 Diet and lipids in early life. In: Barker DJP (ed) Scientific Report No. 8: Infant nutrition and cardiovascular disease. MRC Environmental Epidemiology Unit, Southampton, p 17–19

Lucas A, Blackburn AM, Aynsley-Green A, Adrian TE, Sarson DL, Bloom SR 1980 Breast v. bottle: endocrine responses are different with formula feeding. Lancet 1:1267–1269

Lucas A, Gore S, Cole TJ et al 1984 A multicentre trial on the feeding of low birthweight infants: effects of diet on early growth. Arch Dis Child 59:722–730.

Lucas A, Bloom SR, Aynsley-Green A 1985 Gastrointestinal peptides and the adaptation to extrauterine nutrition. Can J Physiol Pharmacol 63:527–537

Lucas A, Ewing G, Roberts SB, Coward WA 1987 How much energy does the breast-fed infant consume and expend? Br Med J 295:75–77

Lucas A, Brooke OG, Baker BA, Bishop N, Morley R 1989a High alkaline phosphatase activity and growth in preterm neonates. Arch Dis Child 64:902–909

Lucas A, Morley R, Cole TJ et al 1989b Early diet in preterm babies and developmental status in infancy. Arch Dis Child 11:1570–1578

Lucas A, Brooke OG, Morley R, Cole TJ, Bamford MF 1990a Early diet of preterm infants and development of allergic or atopic disease: randomised prospective study. Br Med J 300:837–840

Lucas A, Morley R, Cole TJ et al 1990b Early diet in preterm babies and developmental status at 18 months. Lancet 335:1477–1481

Mesami E, Timcras PS 1982 Normal and abnormal biochemical development of the brain after birth. In: Jones CT (ed) The biochemical development of the fetus and neonate. Elsevier Biomedical Press, New York, p 759–821

Mott GE, Lewis DS, McGill HC Jr 1991 Programming of cholesterol metabolism by breast or formula feeding. In: The childhood environment and adult disease. Wiley, Chichester (Ciba Found Symp 156) p 56–76

Neuringer M, Conner WE, Lin DS, Barstad L, Luck S 1986 Biochemical and functional effects of prenatal and postnatal n3 deficiency on retina and brain of rhesus monkey. Proc Natl Acad Sci USA 83:4021–5

Poskitt EME, Cole TJ 1977 Do fat babies stay fat? Br Med J 1:7–9

Smart J 1977 Early life malnutrition and later learning ability. A critical analysis. In:
    Oliverio A (ed) Genetics, environment and intelligence. Elsevier/North-Holland
    Biomedical Press, p 215–235
Smart J 1986 Undernutrition, learning and memory: review of experimental studies. In:
    Taylor TG, Jenkins NK (eds) Proceedings of XIII International Congress of Nutrition.
    John Libbey, London, p 74–78
Walter T, de Andraca I, Chadud P, Perales CG 1989 Iron deficiency anemia: adverse
    effects on infant psychomotor development. Pediatrics 84:7–17
Whitten CF, Steward RA 1980 The effect of dietary sodium in infancy on blood pressure
    and related factors. Studies of infants fed salted and unsalted diets for five months
    at eight months and eight years of age. Acta Paediatr Scand 279:1–17
Yehuda S, Youdim MBH 1988 Brain iron deficiency: biochemistry and behaviour. In:
    Youdim MBH (ed) Brain iron. Neurochemical and behavioural aspects. Taylor &
    Francis, London

DISCUSSION

Hamosh: What was the breast milk that you used?

Lucas: The banked breast milk in our study was drip breast milk.

Hamosh: This means that the breast-fed babies were deprived of energy, because drip milk has a much lower fat content than suckled or expressed milk.

Lucas: Yes, but at 18 months the difference between babies fed on the term (standard) formula and the preterm formula, which resulted from far less of a difference in nutrient terms than the difference between banked breast milk and the preterm formula, was the greatest difference seen at developmental follow-up. So I don't think the results can be explained by the use of an unusual type of human milk. Our study did give us a chance to compare extremes of nutrient intakes, but much lesser extremes than donor milk versus preterm formula still resulted in major differences in outcome.

Hamosh: Did you measure thyroxine and triiodothyronine (T3) levels in the babies? Recent studies reported at the May 1990 meeting of the Society for Pediatric Research suggest that thyroxine induces higher carnitine levels in muscle of normal as well as hypophysectomized rats (Hug & McGraw 1990). This could explain why breast-fed babies might be leaner, because they could use the energy supplied more efficiently than bottle-fed infants. Breast milk provides the infant with thyroid hormones (Oberkotter 1986).

Lucas: We have monitored T3 levels sequentially in these preterm babies. There is no significant different in T3 secretion between the breast feeders and formula feeders, although there is a slight trend towards higher T3 levels in those fed a preterm rather than standard formula. This contrasts with hormones like growth hormone, prolactin, and gastrointestinal hormones such as motilin, gastric inhibitory peptide, neurotensin and insulin, which are known to be markedly affected by diet in the newborn period. However, we have looked only at T3 levels and not other measures of thyroid status in our preterm babies.

Hamosh: Do you have any carnitine levels?

Lucas: No.

Chandra: In your data on the incidence of allergic reactions you categorized the reactions into eczema, wheezing, and food and drug reactions. There must have been some overlap; for example, in children with eczema, a proportion (perhaps 20%) would also show a reaction to foods. What was the overall effect?

Lucas: The effect of diet on the overall incidence of allergic reactions was very similar to the effect on eczema. In other words, if we divided the babies up by diet group and they showed one or more allergic reactions, there was still a highly significant difference in one direction with a positive family history of atopy and a trend in the opposite direction in the absence of such a history. In fact, in the babies with a family history the incidence of one or more reactions was approaching 50%. This has led us to speculate that premature babies may be a high risk group for allergic disease. There was a very high incidence of eczema and wheezing in these babies. This raises the more general issue of whether there is a relationship between atopy and prematurity.

Chandra: I agree, based on our unpublished information. Among low birth weight infants, the incidence of atopic symptoms was about 60% greater than in full-term infants. This contrasts with the findings of a retrospective study from Britain showing an inverse correlation, namely that among children with atopic eczema, there was a significant lack of preterm infants (David & Ewing 1988). This is rather surprising, because low birth weight infants show increased intestinal absorption of macromolecules as well as disordered immunoregulation, and both these factors would be expected to result in a higher incidence of atopy.

Lucas: Some later reactions seem to be suppressed by early exposure to antigen. Premature babies have an extremely high rate of exposure to drugs; it's common for them to be on 4–10 different drugs for many weeks of their postnatal life. So one might expect them to be at high risk for later drug reactions. But in fact we find that the ill babies, receiving the most intensive care with the greatest exposure to drugs, have the lowest incidence of subsequent drug reactions. We looked at this for penicillin; the incidence of penicillin reactions is lower in babies exposed to this drug in the newborn period than in those not exposed. It is a complicated picture, that some agents will enhance and some will reduce the incidence of later allergic reactions.

Chandra: There might be a differential effect on IgE-mediated reactions, including some reactions to penicillin, compared with other types of allergic reactions (Types II to IV), which may largely be responsible for symptoms such as eczema.

Wadsworth: In very long-term studies of height growth it has been assumed that males experienced more growth retardation than females as a result of adverse environments, exemplified for instance by poor socio-economic circumstances (Bielicki 1986). Can you say any more about sex differences in height and how they may begin?

Lucas: We have only just discovered this marked positive effect of increased nutritional intake on cognitive development in males. We have not identified an interaction between sex and diet for height. Cognitive development stands out, whereas there's no interaction between diet and sex for motor development.

Wadsworth: We found distinct sex differences in height development (Kuh & Wadsworth 1989), and we also find sex differences in the secular trend of height attainment in populations born between the end of the last century and the 1960s (D. J. L. Kuh et al, unpublished work).

Lucas: That is very interesting. I have no idea why males should be more vulnerable to nutritional insults early in life than females.

Meade: What proportion of the babies who were eligible to enter your trial actually did so?

Lucas: This is a completely unselected series of babies. The only thing selected for was birth weight; all babies had to be less than 1850 grams birth weight, to give them a worthwhile time on the trial diets. We excluded a few infants with chromosome disorders that are known to influence growth and development. We approached 926 mothers, asking for informed consent for the baby to take part, and we got 926 acceptances. The only other exclusion was babies who died before they could be randomized, in the first 48 hours of life.

Moxon: I have a couple of methodological questions. First, is the trial blinded, particularly as regards the follow-up observations made on the babies?

Lucas: Yes. It would be impossible to blind the comparison of banked breast milk with formula, because the diets have to be handled completely differently, but the developmental follow-up is done blind. The investigators at follow-up have no idea what the diet groups had been in the neonatal period.

Moxon: So, as a result of the interactions between the mother and the person doing that follow-up, no information as to which group the babies belonged is unintentionally leaked?

Lucas: No. We spent a lot of time making certain that that wouldn't happen.

Richards: Do the mothers know which diet group their babies were in?

Lucas: Yes, but the questioning sequence in the follow-up interview is done in a particular way so that this type of information, if it were to be volunteered at all, wouldn't be given until the major measurements on growth and development had been made. But, by and large, most of the mothers don't talk about their infants' early feeding in the interview.

Moxon: So there is the potential for leakage in the system there, but you feel it is minimal?

Lucas: Ruth Morley, who has done most of the developmental follow-up work, would regard the leakage as non-existent, rather than minimal.

Moxon: Secondly, although you are examining a general hypothesis relating to insults or events that could alter programming during critical periods, you are looking at a multiplicity of potential outcomes, and one would expect some of these, by chance alone, to show differences. Can you give some framework within which to gain a perspective on the observed differences, given the many outcomes which you may have looked at which show no differences?

Lucas: We calculated the size of trial needed for a principal outcome response, namely neurodevelopment, assuming a standard deviation slightly above the population mean, of 17 rather than 15. The trial size was calculated to pick up a pre-set difference.

That was our main outcome response. We have half a dozen subsidiary outcome responses which were pre-specified. I would agree that any further outcomes, which in general I have not focused on in this paper, would be subject to all the criticisms of multiple comparisons. It is extremely difficult to know whether one should use 'punitive' statistics for these, like the Bonferoni procedure. The only honest course is to point out the number of outcome responses that one has looked at, give the significance levels, and allow people to come to their own conclusion. And procedures like the Bonferoni are inappropriate, in any case, because many of these outcome responses are linked, and that kind of system of statistical adjustment would depend on totally unrelated outcome responses. It is highly likely, for example, that weight, length, head circumference, skinfold thickness and all the developmental scores would be all positively correlated with each other.

Lloyd: You have said that the preterm infant is an important model. I would like to suggest that the preterm infant is a very important model *for the preterm infant*, and its subsequent development, but may not be the right model for most infants who are, after all, born at term. The critical period of feeding defined for preterm infants may not be the same if the infant is born at term. I am a little worried that your excellent work is being slipped over as applicable to all infants, and has relevance for early infant nutrition in general.

Lucas: I have made enormous efforts to make it clear that we are talking only about preterm infants. My feeling was that we needed a model where early dietary programming effects are likely to occur. I regard the premature baby as being nearer to fetal life than the term baby, and, therefore, from our understanding that most influences on programming occur earlier rather than later, would be more likely to show an effect of diet. In other words, if we couldn't demonstrate long-term consequences of nutrition in preterm infants, that would be a serious limitation so far as looking at other infants goes. But I agree that if we do confirm long-term effects in premature babies, we have to start from scratch with full-term babies, because—as you say—they may well not be sensitive in the same way as babies born three months too soon. It is a start; that is all I would claim. I agree with the criticism.

Golding: On June Lloyd's point about extrapolating to normal term infants, information from the UK cohort studies confirms the difference between initially breast-fed and bottle-fed babies (Butler & Golding 1986). However, when we analyse the data on the 1970 national cohort (a study of all births in Great Britain in one week of 1970 who have been followed up ever since), we find children with slightly higher mean IQ scores among those who were breast fed (Pollock 1989). The same was found in the 1946 cohort which Mike Wadsworth directs (Rodgers 1978), where there wasn't the bias to breast feeding in the higher social classes that there was in the 1970 cohort. This does seem to indicate that early breast feeding had a long-term effect on mental development.

Lucas: We found the same with our premature babies, that after adjusting for many factors we could not 'expunge' the advantage that babies who had received their own mothers' breast milk had in their later cognitive development. This must mean either that we could not adjust for important parenting factors, which I suspect is quite likely (combined with the problem of non-randomized comparisons); or, alternatively, there is something, or things, in breast milk that promote(s) brain development—long-chain $\omega - 3$ fatty acids, nerve growth factors, other growth factors, and so forth. This is an idea for exploration, but it's interesting that the same data are found for the premature babies and for the term babies studied.

Suomi: I like very much the notion of 'vulnerable' individuals, and I wonder if it is possible that the sorts of effects that you are describing basically are limited to a few individuals who might be genetically 'vulnerable' for one reason or another, rather than to the entire sample? Would you agree with that view? And would you expect different effects in the short-term and long-term consequences, for such vulnerable and non-vulnerable individuals?

Lucas: Firstly, in the study there are some outcome responses that appear to be much more 'robust'; that is, they are affected by early diet in all individuals studied. Thus motor development was very strongly affected by early diet in all the infants we looked at, regardless of whether they were small-for-dates or appropriate-for-dates babies. Other outcomes appear to be much more selectively influenced in high risk groups—presumably these are the outcome responses that are not quite so strongly expressed and are only revealed in those babies in whom the dietary effect is magnified. We have not wanted to get involved in too many different subgroup analyses, at this stage, so we chose the standard dichotomies to look at: sex, whether the infants were growth retarded or not, and whether they were well or ill. But I imagine that the differences observed in the more vulnerable infants in each of these dichotomies are more likely to persist long-term, simply because they are larger and will be more shielded from the effects of other environmental events over the years.

References

Bielicki T 1986 Growth and economic wellbeing. In: Faulkner F, Tanner JM (ed) Human growth. A comprehensive treatise. Plenum Press, New York
Butler NR, Golding J 1986 From birth to five: a study of the health and behaviour of Britain's five year olds. Pergamon Press, Oxford

David TJ, Ewing CI 1988 Atopic eczema and preterm birth. Arch Dis Child 63:435–436
Hug G, McGraw C 1990 Carnitine in serum, liver, muscle and heart of normal and hypophysectomized rats before and after thyroxine treatment. Pediatr Res 27:106A
Kuh DJL, Wadsworth MEJ 1989 Parental height, childhood environment and subsequent adult height in a national birth cohort. Int J Epidemiol 3:663–668
Oberkotter LV 1986 Thyroid hormones in milk. In: Hamosh M, Goldman AS (eds) Human lactation 2: maternal and environmental factors. Plenum Press, New York, p 195–204
Pollock JI 1989 Incidence of breast feeding and its relation to outcomes of health and development. Matern Child Health (suppl) 14(9):6–8
Rodgers B 1978 Feeding in infancy and later ability and attainment: a longitudinal study. Dev Med Child Neurol 20:421–426

Programming of cholesterol metabolism by breast or formula feeding

Glen E. Mott*, Douglas S. Lewis† and Henry C. McGill Jr*†

*Department of Pathology, University of Texas Health Science Center, San Antonio, TX 78284-7750 and †Department of Physiology and Medicine, Southwest Foundation for Biomedical Research, PO Box 28147, San Antonio, TX 78228-0147, USA

Abstract. We tested the hypothesis that breast or formula feeding and cholesterol intake during the neonatal period influence cholesterol metabolism and arterial fatty streaks in young adult baboons. Genetic variation was controlled by randomly assigning half-sib sire progeny to a factorial dietary design. We measured serum cholesterol and lipoprotein cholesterol concentrations enzymically and cholesterol production and bile acid excretion rates isotopically. The bile cholesterol saturation index was calculated from enzymic analyses of cholesterol, bile salt and phospholipid concentrations in gallbladder bile. Breast-fed baboons had higher serum VLDL + LDL cholesterol/HDL cholesterol ratios in the early postweaning period (six months) until adulthood (7–8 years) than formula-fed baboons. In adulthood a high cholesterol diet increased bile acid excretion by approximately 40% in formula-fed baboons but did not significantly increase the bile acid excretion rate among breast-fed animals. Adult baboons breast fed as infants also had an approximately 8% lower cholesterol production rate than formula-fed animals and a 20% higher bile cholesterol saturation index. The level of cholesterol in the infant formulas influenced cholesterol metabolism in adulthood but not serum lipoprotein concentrations. As young adults, breast-fed baboons had more extensive arterial fatty streaks than formula-fed baboons. This difference could be accounted for by differences in the lipoprotein ratios. These results demonstrate that breast and formula feeding differentially modify cholesterol metabolism. This may influence the development of chronic diseases.

1991 The childhood environment and adult disease. Wiley, Chichester (Ciba Foundation Symposium 156) p 56–76

Current studies on the effects of infant diet on plasma lipoproteins and atherosclerosis are based on the hypothesis that the preweaning cholesterol intake may prevent diet-induced hypercholesterolaemia later in life (Fomon 1971, Reiser & Sidelman 1972). Part of this hypothesis is that cholesterol in breast milk has an important role in the development of cholesterol regulatory mechanisms. Several experiments with experimental animals (Reiser & Sidelman 1972, reviewed by Hahn 1989) seemed to support those speculations, but others have not (reviewed by Hamosh & Hamosh 1987, Hamosh 1988). This 'cholesterol hypothesis' has not been tested in humans, but comparisons between breast-fed and formula-fed individuals have not shown consistent postweaning effects on serum cholesterol concentrations (reviewed by Hamosh & Hamosh 1987). Most of the studies with experimental animals and humans measured only serum cholesterol concentrations and did not include measures of lipoproteins, apolipoproteins, or cholesterol metabolic variables. In baboons, the total serum cholesterol concentration was also not influenced by previous breast or formula feeding (Mott et al 1982, 1990). However, we have observed consistent differences between breast and formula feeding for lipoprotein concentrations and measures of cholesterol metabolism in juvenile and adult baboons (Mott et al 1982, 1985, 1990, Lewis et al 1988).

We designed two experiments to determine the effects of early diet on lipid metabolism and subsequent disease processes in baboons. Genetic and dietary factors that are difficult to control in human experiments were carefully controlled in these studies.

Experimental design and results

Experiment 1

This experiment was designed to test whether underfeeding vs overfeeding, and breast vs formula feeding, in infancy influenced the development of obesity and atherosclerosis in young adulthood. Forty-five newborn baboons from six sire families were randomly assigned to breast feeding or to one of three formulas with caloric densities of approximately 41, 68 or 95 kcal/100 g of Similac formula (Ross laboratories, Columbus, OH). At 16 weeks of age all animals were weaned from the infant diets to a high cholesterol, high saturated fat diet (1.7 mg cholesterol/kcal, 40% of energy from fat) and were maintained on this diet until five years of age, when arterial lesions and the degree of adiposity were assessed (Lewis et al 1986, 1988, 1989). Although baboons overfed in infancy had greater adiposity at five years than those normally fed or underfed (Lewis et al 1986, 1989), no significant effects of caloric intake or adiposity on serum lipoproteins or atherosclerosis were observed (data not shown). Because these end-points did not differ among the three formula groups, we combined the data from these groups and compared them to data from the breast-fed group. No differences between breast and formula feeding in total serum cholesterol were observed during the postweaning period. Breast-fed baboons had a lower ratio of very low density lipoprotein plus low density lipoprotein to high density lipoprotein (VLDL + LDL/HDL) cholesterol than formula-fed baboons during the preweaning period (Fig. 1). After weaning and until five years of age, this difference in the VLDL + LDL/HDL cholesterol ratio reversed, so that the lipoprotein ratio was higher among breast-fed than among formula-fed baboons, as shown in Fig. 1 (Lewis et al 1988). The higher VLDL + LDL/HDL ratio was

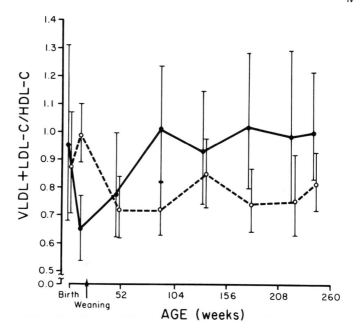

FIG. 1.   Mean VLDL + LDL/HDL cholesterol ratio of baboons ($n = 45$) from birth to five years of age of breast-fed (——) and formula-fed (-----) groups. Bars are 95% confidence limits. (From Lewis et al 1988.)

associated with more extensive arterial fatty streaks in the breast-fed animals than in those formula fed, as shown in Fig. 2.

Experiment 2

In another experiment we tested the hypotheses that (1) the level of cholesterol intake and (2) breast or formula feeding influence cholesterol metabolism in juvenile and adult baboons fed different amounts of dietary cholesterol and types of fat. In this experiment, 80 baboons from six sire families were randomly assigned at birth to breast feeding or to formulas with low, medium or high cholesterol content (approximately 2, 30 or 60 mg cholesterol/dl of formula) (Mott et al 1990). The level of cholesterol in the formula 30 group approximated the cholesterol content of baboon breast milk. After weaning at about 16 weeks of age the baboons in the four diet groups were randomly assigned to one of four adult diets: high cholesterol (1.0 mg cholesterol/kcal) or low cholesterol (0.01 mg/kcal) with approximately 40% of energy from either saturated (P/S = 0.37) or unsaturated (P/S = 2.1) fat (Mott et al 1990). Serum lipoproteins, cholesterol production rate, bile acid excretion rate and cholesterol absorption were measured at intervals and the bile lipid

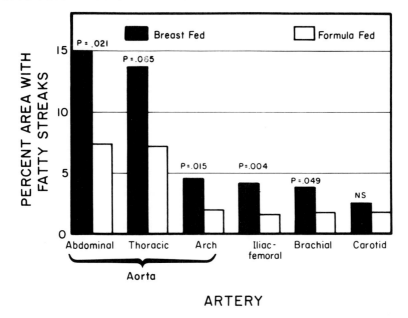

FIG. 2.   Extent of arterial fatty streaks in six arteries at five years of age in breast-fed and formula-fed baboons. (Modified from Lewis et al 1988.)

composition was determined before the animals were necropsied as adults at 7–8 years of age.

During the preweaning period the formula cholesterol content significantly increased the serum cholesterol concentrations from 75 mg/dl at birth to 12-week levels of 152 mg/dl in the high cholesterol formula (formula 60) group, about 130 mg/dl in the breast-fed and medium cholesterol group (formula 30), and only 91 mg/dl in the low cholesterol formula group (formula 2) (Mott et al 1978). After weaning to the adult diets at 16 weeks, dietary cholesterol and saturated fat, as compared to unsaturated fat, increased the serum cholesterol concentration. However, after weaning no differences in total serum cholesterol were observed as a result of the infant diet. The HDL cholesterol concentrations of adult baboons fed formula in infancy were significantly higher ($P = 0.040$) and the VLDL + LDL/HDL cholesterol ratios lower ($P = 0.054$) than in those breast fed (Table 1). Among the three formula groups after weaning we observed no significant differences in HDL cholesterol concentration or in the VLDL + LDL/HDL cholesterol ratio. There were also postweaning differences between breast and formula feeding for the HDL cholesterol response to the type of fat (Fig. 3). The HDL cholesterol concentrations of the three groups of formula-fed baboons were decreased by unsaturated fat compared to saturated fat, but the HDL cholesterol of breast-fed animals was not influenced

TABLE 1   HDL cholesterol and VLDL + LDL/HDL cholesterol ratio after breast or formula feeding in 7–8-year-old baboons

Infant diet	HDL cholesterol (mg/dl)	VLDL + LDL/HDL cholesterol ratio[a]
Breast fed	75.9 (70.2–82.0)	0.631 (0.538–0.741)
Formulas	83.3 (79.0–87.8)	0.526 (0.471–0.586)
Significance	$P = 0.040$	$P = 0.054$

[a]Mean; 95% confidence interval in parenthesis.

by feeding with unsaturated or saturated fat, either during the juvenile period (Mott et al 1982) or in adulthood (Mott et al 1990) (Fig. 3).

We also observed significant differences between breast and formula feeding for several metabolic measures. During the juvenile period, breast-fed baboons when compared to formula-fed baboons had a higher cholesterol absorption rate, lower cholesterol production rate (approximately equivalent to cholesterol turnover rate), and smaller cholesterol mass of the slowly exchanging metabolic compartment (Pool B, which includes muscle, tendons, fat and arteries) (Table 2). Baboons that were breast fed rather than formula fed as infants had lower

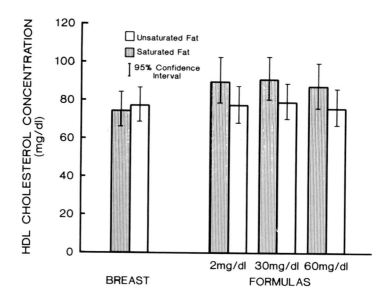

FIG. 3.   Mean serum high density lipoprotein (HDL) cholesterol concentration of adult baboons ($n = 80$) by infant diet and type of adult dietary fat. (From Mott et al 1990.)

TABLE 2 Variables (means) of cholesterol metabolism in baboons during the juvenile period (3.5 years) by breast or formula feeding

Infant diet	Cholesterol absorption (%)	Cholesterol production rate (mg/kg per day)	Cholesterol mass of Pool B (mg/kg)
Breast fed	46.6	33.0	374
Formulas	41.4	36.0	417
Significance	$P < 0.05$	$P < 0.05$	$P < 0.01$

(From Mott et al 1985.)

cholesterol production rates as juveniles (Mott et al 1985) and also as adults (19.5 mg/kg per day vs 21.1 mg/kg per day, $P = 0.014$) (Mott et al 1990). We did not observe significant effects of formula cholesterol content on any postweaning metabolic measure until adulthood. At 7–8 years of age we identified several significant quadratic trends among the three formula groups (Table 3). These quadratic effects were characterized by similar metabolic responses of the high or low cholesterol formula groups. In other words, the cholesterol turnover and bile acid excretion rates for the formula 30 group were higher than those for the other two formula groups. The total exchangeable cholesterol mass (Pools A & B), the cholesterol mass of Pool B, and the half-time of cholesterol in Pool B were lower in the group fed formula 30 than in the other two formula groups (Table 3) (Mott et al 1990; the model of cholesterol metabolism is described elsewhere, in Mott et al 1985). The bile cholesterol saturation index (Redinger & Small 1972) was approximately 20% higher in adult baboons breast fed as infants than in those fed formulas (87.0% vs 72.8%, $P < 0.004$) (Mott et al 1991). No significant differences in the bile cholesterol saturation index were observed among adults fed the three formulas as infants.

In addition to differences between breast and formula feeding for the metabolic variables described above, we observed several interactions of breast or formula feeding with the level of cholesterol in the postweaning diet. No significant interactions of infant formula cholesterol content with the level of postweaning cholesterol intake were detected. Therefore, the data for the formula groups were combined and compared to those for the breast-fed group. Figures 4 and 5 show that as adults, formula-fed baboons dramatically increased their bile acid excretion (turnover) rate and cholesterol production rate in response to dietary cholesterol, whereas breast-fed animals showed minimal response (Mott et al 1990). Similar interactions of the infant diet with adult cholesterol intake were detected during the juvenile period (Mott et al 1985). Significant differences among sire progeny groups were present for serum cholesterol and lipoprotein concentrations, and for measures of cholesterol metabolism (Mott et al 1978, Flow et al 1982, Flow & Mott 1984). Because of the small numbers of animals, we could not estimate interactions between sire progeny group and diet.

TABLE 3 Measures of cholesterol metabolism in adult baboons by infant formula cholesterol content (means and 95% confidence intervals)

Infant diet	Number of animals	Cholesterol turnover rate (mg/kg per day)	Bile acid excretion rate (mg/kg per day)	Total exchangeable cholesterol mass (Pools A & B) (mg/kg)	Cholesterol mass of Pool B (mg/kg)	Half-time of Pool B (days^{-1})
Formula 2	20	16.4 (14.8–18.2)	8.50 (7.47–9.68)	780 (731–832)	516 (471–566)	40.9 (37.9–44.2)
Formula 30	20	19.0 (17.2–20.9)	10.3 (9.12–11.6)	709 (667–753)	438 (402–478)	32.4 (30.1–34.9)
Formula 60	18	16.4 (14.7–18.2)	8.72 (7.61–10.0)	749 (699–802)	479 (434–529)	37.2 (34.2–40.4)
Significance (P) (quadratic trend)		0.022	0.023	0.039	0.018	0.0002

(From Mott et al 1990.)

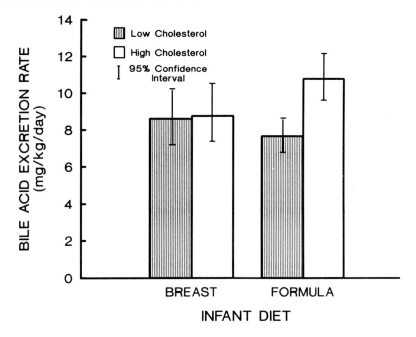

FIG. 4.   Bile acid excretion rate in adult baboons ($n = 80$) by breast or formula feeding and adult dietary cholesterol content. (Mott et al 1990.)

In the second experiment we did not detect statistically significant differences in the extent of arterial lesions between baboons that were breast or formula fed in infancy (Mott et al 1990). However, in the arteries with the most surface involvement—the thoracic and abdominal aortas—breast-fed animals had 50% and 33% more fatty streaks than those fed formula. This result is consistent with the more extensive lesions found among breast-fed baboons in Experiment 1. The absence of statistically significant effects of infant diet on arterial lesions in the second experiment is probably because in this experiment we used a lower cholesterol diet after weaning (1.0 mg/kcal vs 1.7 mg/kcal) and because half of the animals were fed diets low in cholesterol or high in polyunsaturated fat. These diets produced lower overall VLDL + LDL/HDL cholesterol ratios (0.4–0.6) in the second experiment than those in Experiment 1 (0.7–1.0).

Discussion

The factors responsible for the long-term effects of infant diet on cholesterol metabolism are not known. However, several differences in the composition of breast milk and infant formulas are potential mediators of cholesterol regulatory processes. The differences among types of milk and formulas most

Mott et al

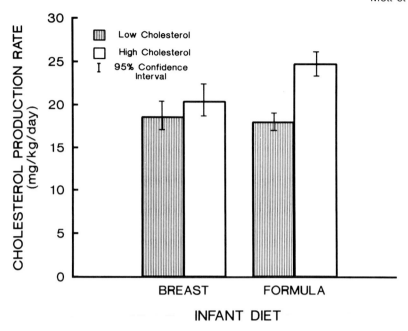

FIG. 5.   Cholesterol production rate in adult baboons ($n = 80$) by breast or formula feeding and adult dietary cholesterol content. (Mott et al 1990.)

frequently studied are the cholesterol and fatty acid content. Reiser's original hypothesis was that dietary cholesterol early in life would 'protect' against diet-induced hypercholesterolaemia in adulthood (Reiser & Sidelman 1972). Our experiments with baboons and most studies with experimental animals do not support this specific aspect of the hypothesis; rather, the results suggest that factors other than cholesterol in milk cause the long-term effects. However, we cannot exclude the possibility that cholesterol and other components of breast milk cause the long-term effects of breast as compared to formula feeding. We propose that a programming effect (metabolic imprinting) by breast feeding or formula feeding results in more efficient utilization of cholesterol by breast-fed baboons. The higher cholesterol absorption and lower cholesterol turnover (excretion) rate among the breast-fed animals are likely to be responsible for the higher VLDL + LDL/HDL cholesterol ratio in breast-fed than in formula-fed animals.

Another nutrient difference between breast milk and most commercial formulas is the generally higher level of saturated fat in breast milk. However, the fatty acid composition of breast milk reflects the type of dietary fat and therefore varies greatly. There is evidence that the fatty acid composition of the preweaning diet influences the serum cholesterol and lipoprotein concentrations during infancy (Carlson et al 1982). There are no reports that the fatty acid composition has any long-term effects on variables of cholesterol metabolism.

Because hormones can 'imprint' metabolism early in life, we speculate that the numerous growth factors and hormones in breast milk (Koldovsky & Thornburg 1987) could mediate the long-term effects of breast as compared to formula feeding on cholesterol metabolism in juvenile and adult baboons. Likely candidates for this role would be thyroid hormones, corticosteroids, and the hormones that regulate their synthesis and secretion. Thyroid hormones and corticosteroids are critical during development and greatly influence cholesterol and lipoprotein metabolism. Differences in plasma gut peptides and other hormones between breast and formula feeding (Lucas et al 1980, 1981) may also influence the development of cholesterol homeostasis.

Acknowledgements

This work was supported by grants HL-15914, HL-19362, HL-28728 and HL-28972 from the National Heart, Lung, and Blood Institute.

References

Carlson SE, DeVoe PW, Barness LA 1982 Effect of infant diets with different polyunsaturated to saturated fat ratios on circulating high-density lipoproteins. J Pediatr Gastroenterol Nutr 1:303–309

Flow BL, Mott GE 1984 Relationship of high density lipoprotein cholesterol to cholesterol metabolism in the baboon (*Papio* sp.). J Lipid Res 25:469–473

Flow BL, Mott GE, Kelley JL 1982 Genetic mediation of lipoprotein cholesterol and apoprotein concentrations in the baboon (*Papio* sp.). Atherosclerosis 43:83–94

Fomon SJ 1971 A pediatrician looks at early nutrition. Bull NY Acad Med 47:569–578

Hahn P 1989 Late effects of early nutrition. In: Subbiah MTR (ed) Atherosclerosis: a pediatric perspective. CRC Press, Boca Raton, FL, p 155–164

Hamosh M 1988 Does infant nutrition affect adiposity and cholesterol levels in the adult? J Pediatr Gastroenterol Nutr 7:10–16

Hamosh M, Hamosh P 1987 Does nutrition in early life have long term metabolic effects? Can animal models be used to predict these effects in the human? In: Goldman AS, Atkinson SA, Hanson LA (eds) Human lactation 3: the effects of human milk on the recipient infant. Plenum Publishing Corporation, New York, p 37–55

Koldovsky O, Thornburg W 1987 Hormones in milk. J Pediatr Gastroenterol Nutr 6:172–196

Lewis DS, Bertrand HA, McMahan CA, McGill HC Jr, Carey KD, Masoro EJ 1986 Preweaning food intake influences the adiposity of young adult baboons. J Clin Invest 78:899–905

Lewis DS, Mott GE, McMahan CA, Masoro EJ, Carey KD, McGill HC Jr 1988 Deferred effects of preweaning diet on atherosclerosis in adolescent baboons. Arteriosclerosis 8:274–280

Lewis DS, Bertrand HA, McMahan CA, McGill HC Jr, Carey KD, Masoro EJ 1989 Influence of preweaning food intake on body composition of young adult baboons. Am J Physiol 257:R1128–R1135

Lucas A, Blackburn AM, Aynsley-Green A, Sarson DL, Adrian TE, Bloom SR 1980 Breast vs bottle: endocrine responses are different with formula feeding. Lancet 1:1267–1269

Lucas A, Boyes S, Bloom SR, Aynsley-Green A 1981 Metabolic and endocrine responses to a milk feed in six-day-old term infants: differences between breast and cow's milk formula feeding. Acta Paediatr Scand 70:195–200

Mott GE, McMahan CA, McGill HC Jr 1978 Diet and sire effects on serum cholesterol and cholesterol absorption in infant baboons (*Papio cynocephalus*). Circ Res 43:364–371

Mott GE, McMahan CA, Kelley JL, Farley CM, McGill HC Jr 1982 Influence of infant and juvenile diets on serum cholesterol, lipoprotein cholesterol, and apolipoprotein concentrations in juvenile baboons. Atherosclerosis 45:191–202

Mott GE, Jackson EM, McMahan CA, Farley CM, McGill HC Jr 1985 Cholesterol metabolism in juvenile baboons. Influence of infant and juvenile diets. Arteriosclerosis 5:347–354

Mott GE, Jackson EM, McMahan CA, McGill HC Jr 1990 Cholesterol metabolism in adult baboons is influenced by infant diet. J Nutr 120:243–251

Mott GE, Jackson EM, McMahan CA 1991 Bile composition of adult baboons is influenced by breast versus formula feeding. J Pediatr Gastroenterol Nutr 12:121–126

Redinger RN, Small DM 1972 Bile composition, bile salt metabolism and gallstones. Arch Intern Med 130:618–630

Reiser R, Sidelman Z 1972 Control of serum cholesterol homeostasis by cholesterol in the milk of the suckling rat. J Nutr 102:1009–1016

DISCUSSION

Barker: One can ask the question in humans of whether breast milk feeding confers high or low rates of ischaemic heart disease and whether that is associated with high or low lipoprotein levels, and the answer will no doubt depend in part on what was put in the bottle with which one is comparing the breast-fed babies. Since that will have been a long time ago, we shall never really know. My guess is that, in our Hertfordshire study, we shall find that the bottle-fed babies will have higher cholesterol levels, consistent with their higher death rates from ischaemic heart disease. But the more interesting question is why it is that men who were breast-fed as babies show an extremely strong inverse relationship between growth (both pre- and postnatal growth) and subsequent ischaemic heart disease death rates. Preliminary results suggest that there is a parallel relation with cholesterol levels. Why should the programming, which you are suggesting is hepatic programming, be related to one's size?

Mott: I don't know of any data that provide a rationale for that. Certainly there are many other factors in the development of atherosclerosis. One factor is tissue susceptibility; individuals develop atherosclerosis at very different rates with identical cholesterol concentrations or identical lipoprotein levels. There are therefore other metabolic characteristics that accelerate or decrease the atherosclerotic process. Possibly, underfeeding or starvation early in life could permanently alter lipoprotein uptake by the endothelium, so that there is greater susceptibility to the development of this disease with a given cholesterol or lipoprotein concentration.

Barker: Is it possible that intrauterine growth retardation would modify the 'programming' of cholesterol after birth?

Mott: Growth retardation could produce some permanent biochemical adaptation which would preserve essential metabolic functions. The physiological price for such adaptation could be an adverse effect on plasma lipid levels or other coronary heart disease risk factors. That is a very speculative hypothesis, but it could be tested in an animal model.

Wood: Both your paper and Alan Lucas's paper have the great methodological virtue that they seem to be mainly passive, so far as the recipient of the nutrition is concerned. But for postnatal feeding, I am not convinced that nutrition is a passive matter. There are wide variations in the behaviour of infants; I am particularly thinking here of June Lloyd's concern about the bigger, full-term infant. I am fairly certain, from clinical experience, that growth is strongly related to food intake and to the behavioural aspects of what the child actually wants. This takes us away from cholesterol studies, but the general topic of the passivity of the model should not escape us.

Richards: Along the same lines, I wonder if, in the baboon studies, behavioural processes could explain some of the results? Whereas the breast-fed infants were with their mothers, the formula-reared ones were getting a lot of human handling. As these animals were being brought up in a fairly stressful environment, and received varying amounts of human handling, this could explain why all the formula-fed groups were very similar, and different from the breast-fed group. We know from a number of animal experiments that early handling can produce long-term differences in responses to stress, and in growth rates.

Mott: This behavioural component exists and is impossible to completely control for in baboons, because we can't get banked baboon breast milk and feed infants with it in a nursery in a similar manner to those receiving formula. The behavioural responses of these formula-fed baboons to humans during the preweaning period are significantly different from those of the breast-fed animals, because those raised in the nursery become attached to humans and are quite tame, compared to those reared by their mothers. Nursery-reared baboons were also socialized in small peer groups for 2–3 hours each day. There were behavioural differences in infancy, but by three and a half years of age the infant diet groups are indistinguishable behaviourally (Coelho & Bramblett 1981, 1982, 1984). The effects on cholesterol metabolism, if they have a behavioural basis, persist long after the behavioural differences have disappeared. Also, there are some programming effects on cholesterol metabolism of the three formulas with different levels of cholesterol. These three formula-fed groups were fed in the same nursery and did not differ behaviourally in either the pre- or postweaning periods.

Lloyd: Was the growth of the baboons receiving breast versus cholesterol milks identical, both in the original feeding period, and subsequently?

Mott: In the second experiment, with different levels of cholesterol in the formulas, there were no differences in growth rates, or in body weights, during the infant feeding period (Mott et al 1978), or in adulthood (G. E. Mott et al, unpublished observations), between breast-fed and formula-fed baboons. In the first experiment I discussed, infant baboons were fed formulas with low or high caloric density which resulted in body weights below or above normal during the preweaning period. The overfed female infant baboons also had higher body weights and adiposity as adults, compared to those normally fed or underfed as infants (Lewis et al 1986, 1989). However, there were no differences in the weight of the adult baboons that were breast fed or fed normal formula in infancy (Lewis et al 1986). Differences in caloric intake during infancy did not influence the lipoprotein concentrations during infancy or adulthood and did not affect arterial lesions at the end of the experiment. There was more variability among the breast-fed infants, because some mothers apparently did not feed their young as efficiently as others; in the nursery, that was much better controlled.

Suomi: There is behavioural evidence from Dr Christopher Coe's laboratory at the University of Wisconsin in Madison suggesting that nursery-reared rhesus monkeys have greatly compromised immune systems, both short-term and long-term, relative to mother-reared animals, when they are reared identically after their first six months of life, which encompasses the weaning period. I shall be presenting data suggesting that there are major behavioural and physiological differences in response to a variety of challenges, in that nursery-reared rhesus monkeys show much greater adrenocortical responses, higher levels of cortisol and ACTH, greater catecholamine turnover, and greater sympathetic nervous system activity, to a variety of challenges. Perhaps some of the interactions that you see with diet, or stress challenges, are also working along the same or similar lines? It goes back to the notion of complex vulnerabilities producing these kinds of interactions.

Mott: Exactly. We don't have data for our animals on immunological variables or hormones at this stage, although a study of hormonal responses to dietary challenge in baboons in later life is under way.

Suomi: I would also like to emphasize the relevance of animal studies to the human situation. It can be argued that the relationship between the diets of our distant ancestors and our current Western diets is about as varied as the relationship between what baboons in the wild eat and what they are given in captivity. Also, an eight-year-old baboon is a relatively young adult; in the wild, baboons can live 25 to 30 years. Long-term studies might yield more compelling data, therefore.

Mott: Yes; these are young adult baboons, equivalent to a 20–25-year-old human.

Barker: There is no overwhelming evidence that the fat intake of the British population has changed much in the past hundred years.

Suomi: I am thinking of tens of thousands of years!

Meade: Dr Mott, to what extent is it now thought that the fatty streak is a precursor of the more advanced atheromatous lesion? From my reading of the literature there is still not a consensus among pathologists that there is necessarily a progression from the fatty streak to a more advanced lesion. I am not convinced that the data on feeding patterns that you showed, although obviously reflected in lipoprotein levels, imply an effect on the onset of the clinical disease that we are concerned with, particularly if the baboon is not naturally given to thrombotic occlusion and clinically manifest thrombotic disease. This is a problem about the animal models for this disease, that the species studied do not on the whole get thrombosis, and therefore don't develop the disease that we are interested in.

Mott: The general opinion from my pathology colleagues is that fatty streaks are precursors of the more advanced fibrous plaques and advanced calcified lesions. At some sites of the arterial wall, fatty streaks are highly correlated with fibrous plaques across time. These fatty streaks appear in humans very early in childhood, which is one of the major arguments for dietary intervention in children.

The evidence, however, is not overwhelming, because it's difficult to prove that a fatty streak at a particular site later becomes a fibrous plaque and, later still, an advanced calcified lesion, resulting in clinical disease. But the general view, which is not universal, is that fatty streaks are precursors of the lesions and result from the deposition of serum lipoproteins, which also occurs in the more advanced stages.

As to whether baboons develop thrombosis, they do so only to a very limited degree in the wild. But if you feed them high fat, high cholesterol diets, they develop atheroma. And many people feel that such diets are the major cause of the high rate of coronary heart disease seen in the West.

Meade: But baboons don't have myocardial infarcts, do they?

Mott: They usually die of other causes, in the wild, but the coronary event is a very late stage in atherogenesis.

Meade: A degree of coronary atherosclerosis is obviously necessary for the development of clinically manifest coronary artery disease. On the other hand, all the men in this room probably have significant coronary atheroma, but not all of us will get clinical coronary disease. Some other process must be superimposed on the atheromatous background that is responsible for the clinical event. Something we need to think about, in the context of this symposium and of the disease in general, is the extent to which the sort of influences that David Barker has been identifying might be operating through another set of mechanisms, including thrombogenesis.

Mott: There is no question that other mechanisms contribute to the development of coronary heart disease. Tissue susceptibility is certainly one of them. I am not suggesting that lipoproteins provide the complete explanation for the pathogenesis, but the final event depends on the progression of atherogenesis throughout life. It may not kill you, but it provides the necessary conditions for complete vascular occlusion.

Meade: The final event is thrombotic. That is now clear for myocardial infarction and for sudden coronary death. It is a different process.

Mott: That thrombotic process is itself due to many other factors. An advanced atherosclerotic plaque may spin off clots due to tissue factors that are released at that site. But that's a very late stage in what began as a fatty streak or fibrous plaque.

Meade: It is what kills you!

Mott: That is right.

Rutter: The effects of early diet on cholesterol metabolism in baboons that you have demonstrated are impressive and clear, Dr Mott, but I remain puzzled by the mechanisms involved in the increased rate of fatty streaks in those who were breast fed. As I understand your results, you have three main findings. First, breast feeding (compared with formula feeding) was associated with a higher lipoprotein ratio, a higher cholesterol absorption rate, a lower cholesterol production rate, and a smaller cholesterol mass of the slowly exchanging metabolic compartment. Very reasonably, you interpret this finding as an indication that infant diet influences later cholesterol metabolism in such a way that there is a *more efficient* utilization of cholesterol. I take it that you mean dietary cholesterol in this connection.

Second, in keeping with this inference, you found that neither bile acid excretion nor cholesterol production was appreciably affected in postweaning breast-fed baboons either by cholesterol levels or by the type of fat in their current diet (see Mott et al 1990). In sharp contrast, in formula-fed baboons, higher dietary cholesterol was associated with a large increase in both.

Thirdly, arterial fatty streaks were *more* prevalent in breast-fed baboons. My question is: what mechanism underlies this last finding? If you suggest that the adult diet constitutes the immediate mediating mechanism, why is there no effect of adult diet in the breast-fed baboons? If breast feeding is associated with a *reduced* responsivity to adult diet, why is there an *increase* in fatty streaks? The finding of a *lack* of effect of adult diet in breast-fed baboons would seem to suggest either that adult diet is irrelevant (presumably suggesting that the cholesterol is endogenously, not exogenously, derived), or that the critical level of dietary cholesterol is extremely low (in that it must be below the 0.01 mg/kcal level in what you described as a 'non-atherogenic diet' in your 1990 paper).

Mott: The 'reduced responsivity' to dietary cholesterol of previously breast-fed adult baboons is in the rate of bile acid formation (equivalent to bile acid turnover rate) and cholesterol production (equivalent to cholesterol turnover rate). These are measures of the principal routes of cholesterol removal from the body. Thus, the lower cholesterol excretion among breast-fed than among formula-fed baboons challenged with a high cholesterol diet could be a result of more efficient hepatic regulation of cholesterol and bile acid secretion, or enhanced cholesterol and bile acid absorption. These effects would be likely to lower hepatic LDL receptor activity. A decreased LDL receptor activity would lower LDL turnover and increase the ratio of plasma LDL/HDL, which accounts for the more extensive arterial fatty streaks of breast-fed baboons.

Although this rationale is consistent with the data so far, there are many gaps in our understanding. We do not know the time sequence of these events or which is the primary event. The molecular mechanisms of gene expression appear to be programmed or imprinted by the feeding regimen. But we do not know how gene expression could be influenced long term (permanently?) by the early environment, or what the mediating factors are. We believe that the high cholesterol adult diet is a way to identify among infant diet groups the underlying metabolic differences that are not readily apparent until the animals are challenged with diet. The effects are only revealed as an interaction between the infant diet (breast or formula feeding) and the adult diet; and we would not have observed many of the differences due to the type of infant diet without the four types of adult diets—two levels of cholesterol, and two different types of fat.

The advantage of this factorial design is that one can partition-out the effects of each of these variables independently. The statistical analysis was done by an analysis of variance with a linear model in which there were five main effects, plus all the two-factor interactions except those including sire. This makes it possible to determine the main effects of breast versus formula feeding, and the interactions with other factors. I didn't discuss the main effects of the other factors, but I should emphasize that the effects of infant diet (breast or formula) are very similar in size to the effects of the adult diets (high or low cholesterol, type of fat), and to the effects of differences among the sire groups. Thus the genetic effects that are superimposed are of the same order of magnitude as the infant diet effects. That is why it's critical to determine what those other effects on cholesterol metabolism are, particularly in a relatively small study with experimental animals or humans, otherwise the effects of the infant diet could be completely masked.

Lucas: When people hear about your data, Dr Mott, they are concerned and upset by the idea that breast-fed infants could be worse off than formula-fed infants in later life. But there is no reason to suppose that breast feeding evolved to confer longevity and freedom from disease in old age in humans, unless there is evidence for there being a selective advantage of postreproductive survival. In primates, is there a good reason for believing that such an advantage exists? If there isn't, perhaps nothing has evolved about breast feeding that is protective in the long term and we should not be surprised to find that it is 'worse' in that respect.

Mott: There seem to be many immediate benefits of breast feeding. In our experiments, breast-fed baboons as adults appear to use cholesterol more 'efficiently': cholesterol turnover (production) rates are lower than in formula-fed baboons, and cholesterol absorption is higher. This may be an advantage during a period (i.e. in infancy) in which cholesterol is a major requirement for the formation of new cell membranes, and other structural components in the cells. Cholesterol is a large molecule that is synthesized in the body from acetate but at considerable energy expense. There would be a definite metabolic advantage in having the most efficient utilization of cholesterol during development, but it may prove a disadvantage later in life if the animal is exposed to a high cholesterol, high fat diet, that raises the circulating lipoprotein concentrations and accelerates atherogenesis. So it's a two-sided coin. The metabolic consequences of breast feeding result in a more favourable situation during growth and development, but later may be detrimental.

Dobbing: For me, breast feeding is incomparably the best, mainly because it was evolved over aeons of time, and we interfere with it at our peril. I find this a more powerful argument than the many other reasons given for breast feeding derived from theoretical considerations (Dobbing 1988). I am not very interested in what 'ought' to be about breast feeding; I am interested in what *is*. In my view, the way to measure this is on the basis of 'outcome' in the child, rather than to rely on any immunological or other theory about what *ought* to be the outcome. An exception is the higher incidence of gastroenteritis in poor conditions when breast feeding is practised; but, even in this case, interpretation of the data requires greater attention to the rules of good epidemiology, and less automatic acceptance of the more simplistic explanations of those who are emotionally, and even politically, involved (Dobbing 1988).

Barker: Alan Lucas presented some challenging ideas about breast feeding conferring a selective advantage for prereproductive survival, but disadvantage for postreproductive survival. Does anyone want to follow that up?

Moxon: An emphasis was missing from the earlier discussion with regard to the importance of reproductive potential. Taken to its ultimate, one would look at these adaptive responses prenatally as well as postnatally, taking into account the argument that higher organisms are merely vehicles for genes and that it is their survival and prevalence which count. Thinking in terms, for example, of the central importance of the adrenal gland, emphasis has been focused on cortisol. One might perhaps focus on sex hormones and the way in which they are important in terms of the ontogeny of sexual organs, fecundity, and the implications for the fate of genes.

Casaer: I would like to ask Dr Mott and Dr Lucas whether they looked for behavioural differences in feeding behaviour. In the Developmental Neurology Project in Leuven we did a series of prospective studies on feeding behaviour in cohorts each of 100 preterm infants. The feeding efficiency was measured by the amount of milk intake per time unit and by the number of pauses the baby makes when drinking 10 ml of milk. The postural score describes the active antigravity posture of the infant at the onset of feeding, the adduction and the flexion of the arms, and the adaptation of the hand posture. This simple description of feeding efficiency and posture relates to the integrity of neonatal nervous system functioning, since it has a good correlation with perinatal neurological conditions and with mental outcome scores at the end of the first year of life (Casaer et al 1982, Daniëls & Casaer 1985). My specific question is: have you observed differences in baboons or human infants under different feeding schedules?

Mott: Feeding behaviour in our baboon studies was only estimated indirectly, from the body weights, as discussed above (p 68).

Lucas: We were looking at infants randomized to different diet groups, but, as Martin Richards pointed out, there may well be interactions between diet and the effects that diet have on other aspects of behaviour. So we are really doing a management trial; we are comparing the effects of giving whole diets, together with the subsequent interactions that they may have. It is certainly true that feeding in a broader sense is changed by what we feed babies on. If we feed babies on banked breast milk, they start to tolerate enteral feeds much sooner than those fed on a preterm formula. There are differences in gastrointestinal tolerance, vomiting rates, and so forth. That is bound to have subsequent effects on maturational changes, which equip the infant for extrauterine feeding. Thus it would be difficult to extricate the effects of diet from the subsequent effects that the choice of diet has on certain behavioural outcomes. Really, we are studying not the effects of diet as such, but the effects of dietary assignment.

Caspi: The issue of behavioural outcomes has been puzzling me, Dr Lucas, since you presented your provocative data (this volume: Lucas 1991). It's not entirely clear that intelligence tests administered in infancy and later in childhood actually 'tap' the same functions, yet you seem to imply that performance on both types of test is related to the early nutritional environment. The question of mechanisms is problematic. I am not sure what mechanism would cause nutritional effects to be so pervasive on what essentially are qualitatively different kinds of behaviour.

Lucas: There are a number of questions here. One is whether early developmental tests predict the results of later tests—whether any developmental measure at 18 months, say, would predict later IQ, or 'achievement'. Firstly, population mean developmental scores at 18 months are more likely to be predictive of later population scores than individual test scores. Furthermore, we have now, using data from our 7–8 year follow-up, correlated the IQ (on the WISC) with 18-month developmental performance tests (for example, Bayley Scales) and found good correlation coefficients of 0.6 or more. But I would admit that developmental tests at 18 months are crude. However, it is generally agreed that in individuals a bad performance on developmental tests at 18 months is more likely to predict bad scores later on, than an average score at 18 months would predict later scores. We did show at 18 months a significant increase in very low scores—what could be called minor neurodevelopmental impairment—in preterm babies who had been fed suboptimally, for example on a standard formula. So I think it likely that our results at 18 months will be related to some extent to scores at 7–8 years, and we shall shortly be able to confirm or refute this.

As to why developmental scores are related to diet, we know that there are major effects of early diet on brain structural development. How those effects relate to functional development in humans is unknown.

Smart: A possible instance of the sort of complication that Martin Richards has suggested is that the better-fed infants receiving your preterm formula might mature more quickly and interact properly with their mothers earlier than the other babies. Perhaps you have information on this?

Lucas: Yes. The better-fed premature babies given preterm formula do get discharged slightly earlier from hospital than those fed on unfortified diets. I accept that there may be important 'knock-on' effects of the type of diet that's given. The worrying thing is that if there are such effects, they may not necessarily be translatable to other countries or other neonatal units, whereas purely nutritional effects would be much more robust and applicable to other situations. However, I would suspect that the nutritional effects are the most important here, rather than the knock-on, interactional effects.

As an example, in the first month of life, a baby fed on a preterm formula increases the weight of its brain by 40%, compared to 20% on banked breast milk. This has been estimated from the relationship we derived between head circumference and brain weight from post mortem samples. So we are seeing major nutritional effects on the structural growth of the brain.

Murray: One complication in studying the brains of psychiatric patients is social class. If you look at brain 'slices' on CT and MRI scans, those of higher educational level and higher social class tend to have a larger area (Pearlson et al 1989); some studies suggest that IQ is correlated with the area of the brain shown on the slice. That would fit in with what you are saying—that individuals in lower social classes are more likely to experience malnutrition during fetal or neonatal life, and malnutrition in turn is likely to be related to brain size.

Lucas: Yes. In fact, the social class factor is taken out of our study; it is a randomized study, so we have an equal social class distribution between the dietary groups. But we haven't yet looked at randomized dietary comparisons within social class strata to see whether we get bigger effects of diet in some social classes. In other words, we haven't looked yet at the interaction between social class and diet. That is an important area to explore.

Suomi: Many other things besides diet go along with social class differences. For example, there is an extensive literature on the styles of interaction between mothers and infants as a function of social class, affecting the amount that they vocalize and the contents of the vocalizations. Also, Dr M. H. Bornstein in my lab. has been looking at the relationship between infant responsivity and amount of maternal stimulation in face-to-face interactions during the first year of life. There is a real interactive effect; too much stimulation to the 'wrong' sort of infant can be problematic, whereas the same amount of stimulation given to a high responsive infant leads to a multiplicity of effects with respect to later intelligence, as measured at four years of age. So there are sources of variance other than strictly nutritional factors that may be showing up in some of these studies.

Murray: Is there any evidence that the extent of stimulation given by mothers will influence crude brain size?

Suomi: In rats there is conclusive evidence that increased stimulation of a variety of sorts early in life results in larger brain size and weight (e.g., the studies of D. Kretch, M. Diamond & M. R. Rosenzweig).

Dobbing: It is not brain size which is influenced so much as very small regions of the brain which are said to show weight changes as a result of a stimulating environment (Katz & Davies 1982). Surely the main determinant of brain size, like the main determinant of liver size in the normal population, is body size? Are social class differences in brain size independent of body size?

Murray: Yes; in our South London samples of psychiatric patients we find an effect of social class, even taking body size into account.

Dobbing: If that is so, you could take the brains out of those of us here and arrange them in order of weight, and from that decide our mental capacity. I do not think you could!

Richards: May I make a comment about gender and evolution? One important message from much of our discussion is that fetal life is a vulnerable stage in our developmental history. The outcome in terms of natural selection is reproduction. The reproductive capacity of females—their ability to nurture a fetus and infant—may be crucial to the later capacity of those children to reproduce themselves. So it makes good sense to protect fetuses, especially the females, because of this intergenerational effect. The fate of the female fetus during pregnancy may influence the quality of the uterus that will carry the next generation. One could take this argument on into postreproductive life; even in Britain, the person next most likely to be caring for a baby, after the mother, is the maternal grandmother. So a grandmother may have an influence on the development of her grandchildren. That is an old and widespread cultural pattern. We need to look at parental care and its outcomes across more than one generation and at the different ways in which it may influence the development of males and females.

References

Casaer P, Daniëls H, Devlieger H, De Cock P, Eggermont E 1982 Feeding behaviour in pre-term infants. Early Hum Dev 7:331–346

Coelho AM Jr, Bramblett CA 1981 Effects of rearing on aggression and subordination in *Papio* monkeys. Am J Primatol 1:401–412

Coelho AM Jr, Bramblett CA 1982 Social play in differentially reared infant and juvenile baboons (*Papio* sp). Am J Primatol 3:153–160

Coelho AM Jr, Bramblett CA 1984 Early rearing experiences and the performance of affinitive and approach behaviour in infant and juvenile baboons. Primates 25:218–224

Daniëls H, Casaer P 1985 Development of arm posture during bottle feeding in preterm infants. Infant Behav Dev 8:241–244

Dobbing J 1988 Medical and scientific commentary on charges made against the infant food industry. In: Dobbing J (ed) Infant feeding. Springer-Verlag, London, p 9–28

Katz HB, Davies CA 1982 The effects of early life undernutrition and subsequent environment on morphological parameters of the rat brain. Behav Brain Res 5:53–64

Lewis DS, Bertrand HA, McMahan CA, McGill HC Jr, Carey KD, Masoro EJ 1986 Preweaning food intake influences the adiposity of young adult baboons. J Clin Invest 78:899–905

Lewis DS, Bertrand HA, McMahan CA, McGill HC Jr, Carey KD, Masoro EJ 1989 Influence of preweaning food intake on body composition of young adult baboons. Am J Physiol 257:R1128–R1135

Lucas A 1991 Programming by early nutrition in man. In: The childhood environment and adult disease. Wiley, Chichester (Ciba Found Symp 156) p 38–55

Mott GE, McMahan CA, McGill HC Jr 1978 Diet and sire effects on serum cholesterol and cholesterol absorption in infant baboons (*Papio cynocephalus*). Circ Res 43:364–371

Mott E, Jackson EM, McMahan CA, McGill HC Jr 1990 Cholesterol metabolism in adult baboons is influenced by infant diet. J Nutr 120:243–251

Pearlson GD, Kim WS, Kubos KL et al 1989 Ventricle–brain ratio, computed tomographic density, and brain area in 50 schizophrenics. Arch Gen Psychiatry 46:690–697

Interactions between early nutrition and the immune system

Ranjit Kumar Chandra

Memorial University of Newfoundland, and Janeway Child Health Centre, St John's, Newfoundland, Canada A1A 1R8

Abstract. The ontogenetic development of the immune system is a well-defined, almost stereotyped event. The anlagen of the human thymus can be distinguished in the third and fourth branchial arches in the sixth week of gestation. Several distinct responses have been observed in the first trimester of fetal development. In addition to genetic factors, environmental influences such as nutrition play an important role in influencing the developing human immune system. Adverse factors that impair fetal growth hinder immunological maturation as well. These include maternal malnutrition, smoking, alcohol and other substance abuse, placental insufficiency and infection. The immuno-competence of low birth weight infants is compromised; those who are small for gestation show persistent immunological impairment for several months, even years. Prolonged effects on immune responses can be seen in animal models of fetal malnutrition. A second area where interactions occur between dietary factors and the immune system is the IgE-mediated allergic response. Those with a family history of atopy are at high risk of developing allergic disease in late childhood and adult life. The enormous costs of atopic disease in terms of health management and physical and emotional isolation have led to attempts at prevention. These strategies include restriction of the mother's diet during pregnancy and lactation to exclude common allergenic foods and prolonged exclusive breast feeding. Casein or whey hydrolysate formulas are advisable in those not breast fed.

1991 The childhood environment and adult disease. Wiley, Chichester (Ciba Foundation Symposium 156) p 77–92

Diet and health are intimately linked to each other. This relationship is of considerable importance during early life, including fetal life and early infancy. These considerations apply to the immune system as well. Two aspects of interactions between nutrition and the immune system are included in this selective review: firstly, the influence of fetal growth retardation on immune responses of the newborn and young infant; and, secondly, the role of diet in the occurrence of atopic disease among infants at high risk.

Development of the immune system

Much of our current knowledge of the development of the immune system is derived from comparative studies in a number of mammalian and non-mammalian species, from amphibians to non-human primates. The ontogenetic differentiation steps are largely genetically regulated but some environmental influences, such as nutrition and infection, can influence this process. The topic has been reviewed (Chandra & Matsumura 1979, Hayward 1981, Miyawaki et al 1981, Haynes et al 1989). Differentiation and maturation of various types of cells that participate in the immune response progress at different rates. Moreover, antigenic differentiation may precede functional development by a substantial time period. Stem cells seen in the fetal liver and bone marrow give rise to B and T cells, as also to erythroid, myeloid and megakaryocytic lineages. Stem cell maturation occurs *pari passu* with a reduction in the capacity to generate distinct cell types.

T cell development

In man, the thymus is derived embryonically from the third and fourth branchial arches. This explains the frequent occurrence together of abnormalities involving both the thymus and the parathyroids, the former manifesting as frequent infections and the latter as persistent neonatal hypocalcaemia. The earliest histological evidence of the thymus is seen at about the sixth week of gestation. The first sign of lymphoid population in the epithelial anlagen is observed at about 8–9 weeks. This is soon followed by lobular organization of the organ and then the clear demarcation between the cortex and the medulla with several epithelial structures, the Hassall's corpuscles. Age-specific characteristic histological features parallel the differentiation of cell types bearing specific antigens on their surface; some antigens are lineage specific, being expressed by all T cells but by no other cell types. Other types of antigen identify the stage of development and the principal site of localization. Still others characterize functionally distinct subsets, for example CD4[+] helper cells and CD8[+] suppressor cells.

There are limited data on functional correlates of ontogenetic development. Responses to mitogens are detected at 10–11 weeks of gestation, whereas alloantigen-specific cytotoxic responses are observed from 16 weeks onwards. Natural killer (NK) cell activity can be demonstrated beyond 9–10 weeks. Newborns have increased suppressor cell activity, which is radiosensitive and differs in several respects from suppression mediated by adult T cells (Lawton 1984). There is near-normal helper T cell activity for IgM antibody responses but reduced helper activity for IgG antibody responses.

B cell development

An early cell type of the B lineage, the pre-B cell, is characterized by the expression of µ chains without any light chains or membrane-associated immunoglobulin (Ig). Clusters of these cells can be seen in the human fetal liver at seven weeks of gestation. By the 24th week, the predominant site of B cells shifts to the bone marrow. The rapidly dividing B cells give rise to the smaller, non-dividing B cells that express light chains and, later on, surface Ig. It is probable that κ light chains arise before λ chains. An infinite number of rearrangements of V_H and V_L genes permits the generation of genetic diversity at the expense of enormous cell wastage.

Immunocompetence of low birth weight infants

The world-wide incidence of low birth weight, defined as weight less than 2500 g, varies considerably from one population group to another, from 8% in some industrialized countries to as many as 41% in some developing countries of Africa. In the former countries the majority are preterm appropriate for gestational age (AGA), whereas in the latter, the majority are small for gestational age (SGA). The aetiology of fetal growth retardation includes maternal malnutrition and infection, hypertension, toxaemia, smoking, substance abuse, and 'placental insufficiency'. More than one factor is often at work.

Low birth weight is associated with a higher mortality. Whereas the overall proportion of infants who died or were handicapped is similar in AGA and SGA groups, the former are at higher risk of death in the immediate postnatal period while the latter are at higher risk of morbidity in the first year of life. Infection is one of the recognized causes of increased illness in SGA infants. In our experience, upper and lower respiratory tract infections are three times more frequent in SGA infants than in AGA infants. It appears that the morbidity pattern in the SGA group has a bimodal distribution, about two-thirds showing a near-normal rate of illness, comparable to that of healthy full-term infants, whereas one-third have an increased illness rate, almost three times that of the full-term infants (Fig. 1). The SGA group is also at risk of developing infection with opportunistic microorganisms, such as *Pneumocystis carinii*, as observed also in postnatal malnutrition.

Cell-mediated immunity

SGA infants show atrophy of the thymus and prolonged impairment of cell-mediated immunity. These general findings have been seen in infants as well as in laboratory animals (Chandra 1975a,b, 1986, Moscatelli et al 1976). In animal models of intrauterine nutritional deficiency, protein-energy malnutrition (Fig. 2) as well as deprivation of selected nutrients (Fig. 3) results in reduced immune responses in the offspring. Briefly, the number of rosette-forming mature T lymphocytes is reduced and their response to mitogens is decreased. Delayed cutaneous hypersensitivity to a variety of microbial recall antigens as

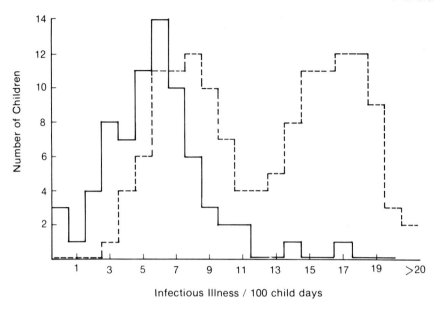

FIG. 1.   Bimodal distribution of morbidity due to infection in small-for-gestational age low birth weight infants (interrupted line) and full-term infants (continuous line) in the first 12 months of life.

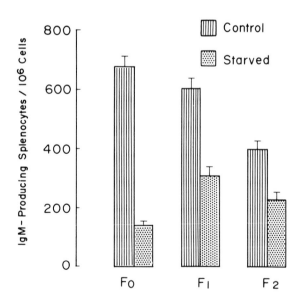

FIG. 2.   IgM antibody-producing spleen cells in partially starved and control rats and their first- and second-generation offspring who were given free access to food.

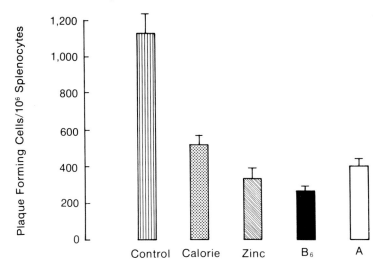

FIG. 3.   IgM antibody plaques in the nine-week-old offspring of adult female rats deprived of calories, zinc, vitamin B$_6$ or vitamin A for four weeks before and during pregnancy.

well as to the strong chemical sensitizer 2,4-dinitrochlorobenzene is impaired. Serum thymic factor activity is reduced in SGA infants tested at one month of age or later (Chandra 1981). In contrast to AGA low birth weight infants, who recover immunologically by about 2–3 months of age, SGA infants continue to exhibit impaired cell-mediated immune responses for several months or even years (Chandra et al 1977). This is particularly true of those infants whose weight-for-height is less than 80% of standard. The prolonged immunosuppression in some SGA infants correlates with clinical experience of infectious illness (Table 1) (Chandra 1986), and thus may have considerable biological significance.

TABLE 1   **Immunological findings in small-for-gestational-age low birth weight infants**

Immunological dysfunction	Group I	Group II
Illness days/100 child days (mean)	6	17*
Impaired cell-mediated immunity	16	14*
Decreased IgG	4	4
Decreased IgG2 level	5	7*
Reduced opsonic activity of plasma	7	9*

41 infants were divided into two groups: those who were relatively asymptomatic (Group I, $n=25$) and those often ill with infection (Group II, $n=16$). Data are shown as the number of infants with the abnormality, except for illness days, which are shown as mean.
*Significantly different from Group I.

Phagocyte function is deranged in low birth weight infants. There is a slight reduction in the ingestion of particulate matter and a significant reduction both in metabolic activity and in bactericidal capacity (Chandra 1975a).

Humoral immunity

IgG from the mother acquired through placental transfer is the principal immunoglobulin in cord blood. The half-life of IgG is 21 days and thus, all infants show physiological hypoimmunoglobulinaemia between three and five months of age. This is pronounced and prolonged in low birth weight infants (Chandra 1975a), because their level of IgG at birth is significantly lower than that of full-term infants. There is a progressive rise in IgG concentration with gestational age and birth weight, especially in infants below 2500 g (Chandra & Matsumura 1979). All four subclasses of IgG are detected in fetal sera as early as 16 weeks of gestation, the bulk consisting of IgG1 (Chandra 1975c). In SGA low birth weight infants the cord blood levels of IgG1 are reduced much more than those of the other subclasses. Thus the infant:maternal ratio is significantly low for IgG1 but not for IgG2 (Table 2). The number of immunoglobulin-producing cells and the amount of immunoglobulin secreted is decreased in SGA infants who are symptomatic—that is, those who have recurrent infections. In the second year of life, SGA infants show a marked reduction in IgG2 levels and often show infections with organisms having a polysaccharide capsule.

Maternal and infant diets and atopic disease

There are two main reasons for renewed interest in the prevention of atopic or allergic disorders. Firstly, the cumulative incidence and prevalence of atopy and of atopic disorders has increased in the last 10–15 years in many industrialized countries. For example, in 12–16-year-old schoolchildren the prevalence of positive skin tests to a battery of inhalant and food antigens has gone up by almost 50% in the last 12 years in Newfoundland. The prevalence of atopic eczema has increased by the same magnitude. The prevalence of bronchial hyperresponsiveness, as judged by a reduction in peak expiratory flow

TABLE 2   Cord blood:maternal blood ratio of immunoglobulin levels in healthy and SGA infants

Ig class	Healthy infants (n = 21)	SGA infants (n = 12)	Significance
IgG1	1.2 ± 0.2[a]	0.7 ± 0.2	P < 0.01
IgG2	0.6 ± 0.1	0.5 ± 0.1	NS

[a]Standard deviation.

after a defined period of moderate physical exercise, has increased by 100%, and the prevalence of asthma has increased by almost three-fold in the last 12 years. The reasons for these changes are not clear.

Secondly, atopic diseases are associated with considerable costs. Many consequences of allergic disorders are immeasurable, such as emotional stress, sociocultural isolation, and school days lost. Others, such as visits to physicians, hospitalization, medicines, special diets, or home renovations, can be measured in terms of monetary values. We have estimated that in Newfoundland, with a child population of 214 368, the annual cost of the management of children with atopic disease is approximately 5.8 million Canadian dollars. Any proposed plan of intervention and prevention should take these costs into consideration.

Identification of the high risk infant

Attempts at the prevention of atopic disease must, of necessity, be targeted to those infants who are at high risk of developing these disorders. There are three recognized methods of identifying such infants at birth. Firstly, a positive family history of atopy among first-degree relatives is a simple and useful indication of predisposition to allergic disease. If one parent is affected, the risk in each offspring is about 37%; if both parents are affected, it is about 62%. In as many as 30% of infants with a definite atopic disease such as eczema or asthma, the family history is negative. Thus attempts have been made to identify other methods and to assess their positive and negative predictive value. Secondly, an elevated cord blood IgE concentration indicates a high risk of developing atopic disease (Croner et al 1982, Chandra et al 1985). Various investigators have used different cut-off points, from 0.5 U per ml to 1.3 U per ml; the former would be very sensitive but would have low positive predictive value. Thirdly, a reduced number of CD8$^+$ suppressor T cells is indicative of high risk (Chandra & Baker 1983). This may be due to the immunoregulatory role of these cells; in situations where there is a reduction in the number and/or function of CD8$^+$ cells, IgE production and levels are increased.

Study design considerations

Intervention and prevention studies in the field of maternal and infant diet and atopic disease are often plagued with problems of faulty design. This makes it difficult to evaluate and compare the results obtained. Unfortunately, some who have attempted a meta-analysis of published investigations often fail to give weight to various criteria of evaluation (Kramer 1988); some aspects of the studies are much more critical than others. For example, studies that depend upon long-term parental recall (Kramer & Moroz 1981, Midwinter et al 1987) are unlikely to yield valid reproducible conclusions.

The important aspects of an adequate design and analysis of studies of the long-term health effects of infant feeding have been reviewed (Chandra 1989, 1990). Briefly, these should include prospective rather than retrospective investigation, stratification of study subjects into high risk and low risk groups, noting the duration and exclusivity of infant feeding, blinding of parents and of observers, objective criteria of assessment and of grading severity of illness, and controlling for confounding variables. It may be true that the perfect study can probably never be done in free-living human populations, but this is no reason for not trying to make the study design and analysis as nearly perfect as possible.

Effect of exclusive breast feeding

There are at least five possible reasons for expecting breast-fed infants to show a reduced occurrence of atopic disease. Firstly, breast-fed infants are less exposed to 'foreign' dietary antigens. Secondly, human milk is postulated to contain factors that mature the intestinal mucosa, thereby allowing early 'closure' of macromolecular absorption. Thirdly, by reducing the incidence of infection and altering the gut microflora that can act as an adjuvant for ingested food proteins, the possibility of sensitization may be decreased. Fourthly, human milk has anti-inflammatory properties that will also decrease macromolecular uptake. Finally, the presence of various cytokines in human milk may play an important role in modulating the development of allergic disease.

Our cumulative experience of exclusive breast feeding and atopic disease is shown in Table 3. Three points should be emphasized. Firstly, the distinct advantages of breast feeding are seen mainly in those infants who are at high risk of development of atopic disease because of family and/or cord blood findings. Secondly, the duration of exclusive breast feeding correlates with outcome; infants breast fed for over four months have a reduced incidence of atopic eczema compared with the group breast fed for less than four months. Thirdly, even among infants fed exclusively at the breast for prolonged periods of time, a substantial number will develop atopic disease. This has led to the investigation of maternal diet during pregnancy and lactation.

TABLE 3   Incidence of atopic eczema up to 36 months of age in breast-fed and formula-fed infants

Group	Low risk infants	High risk infants
Cows' milk formula	3.8%	48.6%
Breast fed		
<4 months	2.8%	32.6%
>4 months	1.6%	22.1%

We have examined the influence of the almost complete exclusion of milk and dairy products, eggs, fish and peanuts from the mother's diet *during pregnancy* on the incidence of atopic eczema in high risk infants. Mothers were recruited before 10 weeks of pregnancy; this is critical, because the fetus is capable of producing IgE after this age. Among those who were breast fed, maternal dietary precautions during pregnancy showed a significant beneficial effect in terms of both the incidence and severity of atopic eczema (Fig. 4) (Chandra et al 1986). Interestingly, the benefit was observed even among formula-fed infants, which may indicate that intrauterine sensitization may have been a factor in all study infants. In another study, maternal diet did not show an effect on the occurrence of atopic disease in the infant (Lilja et al 1988). However, it is important to distinguish the different study designs. In the Swedish study, only two food items were restricted, namely milk and egg, and that too in the last trimester only. From ontogenetic considerations and the possibility of several food antigens being important in this regard, it is not unexpected to find differences in the outcome of the two investigations. Clearly, we need more work in this area.

At least two studies have documented the beneficial preventive effect of maternal dietary precautions *during lactation* among breast-fed high risk infants (Chandra et al 1989a, Hattevig et al 1989). In both studies a reduction in the incidence of atopic symptoms and signs was observed. Our results are summarized in Fig. 5. In Hattevig et al (1989) the differences between study and control groups were significant at three and six months, but not at nine and 12 months. The latter may be the result of a type II statistical error, since a

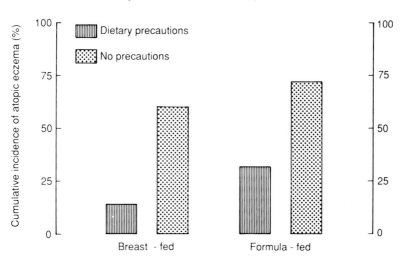

FIG. 4.   Influence of maternal dietary precautions during pregnancy on the incidence of atopic eczema in high risk infants, either breast fed or formula fed. Data up to 18 months are shown.

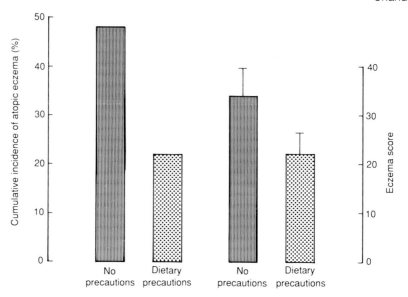

FIG. 5. Influence of maternal dietary precautions during lactation on the incidence and severity of atopic eczema among high risk breast-fed infants. Data up to 18 months of age are shown.

small increase in sample size would have resulted in a statistically significant difference even at nine and 12 months of age. It is interesting that a subsequent follow-up at 18 months has again revealed a statistically significantly reduced incidence of atopic disease in infants whose mothers took dietary precautions while breast feeding (N.I.-M. Kjellman, personal communication).

Effects of various infant formulas

In infants not breast fed, or those who need supplementation, the choice of an appropriate infant formula is important. In two separate studies we have observed a significant preventive effect with the use of a cows' milk protein

TABLE 4   Incidence of atopic eczema up to 18 months of age in non-breast-fed infants

	Infants with eczema	
Group	Number	Mean score
Cows' milk ($n = 40$)	28	55
Soy milk ($n = 41$)	26	56
Casein hydrolysate ($n = 43$)	9**	31*

*$P < 0.05$; **$P < 0.005$.

hydrolysate formula, both casein hydrolysate (Nutramigen, Mead Johnson) (Table 4) (Chandra et al 1989a) and whey hydrolysate (Good Start, Carnation) (Fig. 6) (Chandra et al 1989b) being effective among high risk infants. Considerations of tolerance, acceptance, taste and cost may dictate the choice of one over the other; at present, the casein hydrolysate formula is 2–3 times more expensive than the whey hydrolysate formula. Soy formula feeding was not associated with any protective effect. Although there is some controversy over the usefulness of soy formula in this area, the results of several recent studies indicate little or no benefit (Miskelly et al 1988).

Concluding remarks

There are important interactions between early fetal and infant nutrition and subsequent health outcome in later life. Fetal malnutrition, epitomized by small-for-gestational-age, low birth weight infants, results in pronounced and prolonged impairment of several aspects of immunocompetence. This is associated with an increased incidence of infection in the first few years of life. A second major aspect of the role of early nutrition on health outcome is atopic disease. Exclusive breast feeding, particularly if it is prolonged beyond four months, is associated with a reduced occurrence of allergic disease. The benefit is enhanced by selective exclusion of common allergenic foods from the mother's diet during lactation, and probably also during pregnancy. In the case of infants

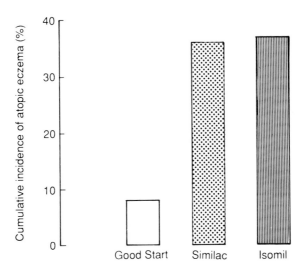

FIG. 6.   Effect of feeding cows' milk whey hydrolysate (Good Start), conventional cows' milk (Similac) or soy (Isomil) formulas on the incidence of atopic eczema in high risk infants. Data up to six months of age are shown.

who are not breast fed or require a supplement, the choice of a hydrolysate formula is distinctly better than a conventional cows' milk formula or a soy formula.

Acknowledgements

Our studies in the area of nutrition and immunity have been supported by the World Health Organization, the Medical Research Council of Canada, the National Health Research Development Program of Health and Welfare Canada, the Indian Council of Medical Research, Sandoz Nutrition, and Abbott Laboratories. Studies on atopic disease have been supported by the National Health Research Development Program of Health and Welfare Canada, Mead Johnson, Ross, and Carnation Nutritional Products.

References

Chandra RK 1975a Fetal malnutrition and postnatal immunocompetence. Am J Dis Child 129:450–455
Chandra RK 1975b Antibody formation in first and second generation offspring of nutritionally deprived rats. Science (Wash DC) 190:189–190
Chandra RK 1975c Levels of IgG subclasses, IgA, IgM and tetanus antitoxin in paired maternal and foetal sera: findings in healthy pregnancy and placental insufficiency. In: Hemmings WA (ed) Maternofoetal transmission of immunoglobulins. Cambridge University Press, London, p 77–90
Chandra RK 1981 Serum thymic hormone activity and cell-mediated immunity in healthy neonates, preterm infants and small-for-gestation age infants. Pediatrics 67:407–411
Chandra RK 1986 Serum levels and synthesis of IgG subclasses in small-for-gestation low birth weight infants and in patients with selective IgA deficiency. Monogr Allergy 20:90–99
Chandra RK 1989 Long-term health effects of infant feeding. Nutr Res 9:1–4
Chandra RK 1990 Long-term health consequences of early infant feeding. In: Atkinson SA, Hanson LA, Chandra RK (eds) Breastfeeding, nutrition, infection and infant growth in developed and emerging countries. ARTS Biomedical Publishers, St John's, Newfoundland, p 47–55
Chandra RK, Baker M 1983 Numerical and functional deficiency of suppressor T cells precedes development of atopic eczema. Lancet 2:1393–1394
Chandra RK, Matsumura T 1979 Ontogenetic development of immune system and effects of fetal growth retardation. J Perinat Med 7:279–287
Chandra RK, Ali S, Kutty KM, Chandra S 1977 Thymus-dependent lymphocytes and delayed hypersensitivity in low birth weight infants. Biol Neonate 31:15–18
Chandra RK, Puri S, Cheema PS 1985 Predictive value of cord blood IgE in the development of atopic disease and role of breast feeding in its prevention. Clin Allergy 15:517–522
Chandra RK, Puri S, Suraiya C, Cheema PS 1986 Influence of maternal food antigen avoidance during pregnancy and lactation on incidence of atopic eczema in infants. Clin Allergy 16:565–569
Chandra RK, Puri S, Hamed A 1989a Influence of maternal diet during lactation and use of formula feeds on development of atopic eczema in high risk infants. Br Med J 299:228–230
Chandra RK, Singh GK, Shridhara B 1989b Effect of feeding whey hydrolysate, soy and conventional cow milk formulas on incidence of atopic disease in high risk infants. Ann Allergy 63:102–106

Croner S, Kjellman NI-M, Eriksson B 1982 IgE screening in 1701 newborn infants and the development of atopic diseases during infancy. Arch Dis Child 57:364–368

Hattevig G, Kjellman B, Sigurs N 1989 The effect of maternal avoidance of eggs, cows' milk and fish during lactation upon allergic manifestations in infants. Clin Allergy 19:27–32

Haynes BF, Denning SM, Singer KH, Kurtzberg J 1989 Ontogeny of T-cell precursors: a model for the initial stages of human T-cell development. Immunol Today 10:87–91

Hayward AR 1981 Development of lymphocyte responses and interactions in the human foetus and newborn. Immunol Rev 57:39–60

Kramer MS 1988 Does breast feeding help protect against atopic disease? Biology, methodology, and a golden jubilee of controversy. J Pediatr 112:181–190.

Kramer MS, Moroz B 1981 Do breast feeding and delayed introduction of solid foods protect against subsequent atopic eczema? J Pediatr 98:546–550

Lawton AR 1984 Ontogeny of the immune system. In: Ogra PL (ed) Neonatal infections. Grune & Stratton, New York, p 3–20

Lilja G, Dannaeus A, Fälth-Magnusson K et al 1988 Immune response of the atopic woman and foetus. Effects of high- and low-dose food allergen intake during late pregnancy. Clin Allergy 18:131–142

Midwinter RE, Morris AF, Colley JRT 1987 Infant feeding and atopy. Arch Dis Child 62:965–967

Miskelly FG, Burr ML, Vaughan-Williams E 1988 Infant feeding and allergy. Arch Dis Child 63:388–393

Miyawaki T, Moriya N, Nagaoki T 1981 Maturation of B-cell differentiation ability and T cell regulatory function in infancy and childhood. Immunol Rev 57:61–99

Moscatelli P, Bricarelli FG, Piccinini A, Tomatis C, Dufour MA 1976 Defective immunocompetence in foetal malnutrition. Helv Paediatr Acta 31:241–247

DISCUSSION

Wood: The cytokine that is most involved in the induction of IgE synthesis is IL-4 (interleukin 4) (Chretien et al 1990). Have you any information about whether any of the groups of subjects in which the development of atopy can be influenced are less adept at producing this cytokine? Going further, and returning to the concept of programming, is there a way of doing something in early life, or before birth, which could permanently influence, and down-regulate, IL-4 production?

Chandra: We have done only a one-point analysis in our study of the different cytokines. In fact, not only is IL-4 production reduced in the breast-fed infants, but IL-2 production is increased. How these two changes might mediate some of the results that we are seeing is not clear, but this is one of the mechanisms that we have to consider.

Barker: Is there any other experimental evidence that is consistent with your findings of an effect on antibody production down to the F2 generation, after starving female rats?

Chandra: There are published results on other organ systems. Zamenhof et al (1968) measured brain DNA content in the *ad libitum*-fed offspring of starved rats. There was a reduction in the F1 and F2 generations.

Barker: How does this operate?

Chandra: There are several mechanisms to consider. We know that growth, as judged by the birth weight of these rat pups, is reduced. Low birth weight is generally associated with compromised immunocompetence, which may never recover completely.

Secondly, subclinical infection could have been a factor. We have not looked at this in our animal experiments, but intrauterine infection such as congenital rubella can have almost a permanent effect on the immunocompetence of the offspring.

The third factor would be stress hormonal factors which can cross the placenta. The stress of starvation causes a number of hormonal changes; free cortisol, not bound to protein, is increased and can produce thymic atrophy and immunosuppression in general. Also, increased levels of α-fetoprotein, seen in fetal malnutrition, may result in immunosuppression.

Barker: But it's not just the daughters, but the granddaughters, of the starved rats that show impaired antibody production.

Chandra: The effect is less marked in the second-generation offspring. Similar results have been seen after maternal deprivation of vitamin B_6 and zinc.

Smart: The results you described on undernourished rats and mice, and in growth-retarded babies, all seem to indicate that *prenatal* growth restriction has a lasting effect on the immune system. Is there any indication of an effect of *postnatal* undernutrition, or is gestation the sensitive period for this?

Chandra: We do see very profound effects of early postnatal malnutrition on immune responses. Most of the effects can be easily corrected by appropriate dietary supplementation. In human infants, the first six months after birth may be a critical period. Infants in orphanages, found to be malnourished at the age of six months, were studied one year later and showed reduced immunological responses (Dutz et al 1976). However, beyond six months of age it seems to be quite easy to reverse malnutrition-induced immunological changes.

Martyn: You showed that IgE concentration in cord blood was an important predictor of allergy later. Does IgE cross the placenta? And what are the major determinants of the concentration in cord blood?

Chandra: IgE does not cross the placenta. When we find an elevated level we have to ensure that it is not the result of maternal blood contamination.

Why is it elevated? There are two possibilities: first, the infant is sensitized *in utero* with a known or unknown antigen; and secondly, he or she has the potential, perhaps because of abnormalities in immunoregulatory mechanisms, to increase the IgE levels. Specific IgE antibodies are rare in cord blood; among 100 infants with elevated cord IgE levels, only 10 or so would have specific IgE antibodies. What antigens are involved in the other infants with elevated cord blood levels of IgE, we don't know.

The predictive value of total IgE is certainly well established. The best studies are from Dr Kjellman's group, who have followed high risk infants for about 10 years and find an almost 80% incidence of atopic disease if cord blood IgE was elevated (Björkstén & Kjellman 1987).

Martyn: And if you compare your two groups of mothers, do you find lower cord IgE concentrations in the group whose diet was restricted?

Chandra: We looked at this, but did not find any difference in the cord blood IgE levels of the two groups. This may be a reflection of the infrequent occurrence of specific IgE antibodies in cord blood.

Moxon: As a more general point, the immune system is basically a contingency system, which is able to adapt to meet different antigenic challenges in a unique way. The immune cells achieve diversity by rearrangements of their genome and by somatic mutation. Monozygotic twins, which are genetically identical, are not identical with respect to their maturing immune systems because of random events occurring in the development of immune cell functions. This thought might be useful when we are considering this rather special system.

In relation to the ontogeny of the immune system and the timing of exposure to particular antigens, what is the effect of an earlier exposure of the premature infant to the multiple antigens associated with birth and subsequent events?

Chandra: There is very little information on this. The preterm infant is able to respond quite adequately, within six or eight weeks of birth, to immunization against tetanus toxoid. Antibody levels comparable to those found in full-term infants have been observed. Unfortunately, large numbers of preterm or small-for-gestational age (SGA) infants have yet to be examined for the protective efficacy of these immunization procedures. In addition, some types of immune responses, such as the secretory IgA antibody response, may be lower than normal, even though the serum antibody response is normal. When such infants encounter natural infection, there may be a severe Arthus reaction. In Africa, measles infection in SGA low birth weight infants who have been given the measles vaccine in the first year of life may produce a severe illness in some of them. Although this situation has not been looked at in terms of a dichotomy between serum and secretory antibody response, I suspect that this may be one of the explanations.

Thornburg: You showed that the fetal:maternal ratio of IgG concentrations is reduced in SGA infants. What is the reason for that? Could it be a placental defect, with a decreased number of Fc receptors, or perhaps a difference in the clearance of IgG?

Chandra: It's not clear whether either of these two factors might be important, but it would be worth looking at. We know that the transfer of IgG is an active process, because even hypogammaglobulinaemic mothers have babies with near-normal IgG levels, and cord blood levels are usually higher than maternal levels, so it's not a passive transfer.

Thornburg: It has been shown that some amino acid concentrations in cord blood are reduced in intrauterine growth retardation (Cetin et al 1988). That may fit in with the same sort of placental defect.

Chandra: The transfer of various specific antibodies may occur at different rates in the low birth weight infant. We looked at antibodies to tetanus antitoxin and found a good correlation with the extent of the weight deficit (Chandra 1975).

Barker: Dr Thornburg, are you saying that in SGA or IUGR babies, there are specific placental lesions which are responsible for the baby being small?

Thornburg: It appears that in these babies there are either fewer receptors for certain substances being actively transported, or the receptors transport less efficiently, so that there is an overall defect in the rate at which receptor-mediated transport is taking place. I am not saying that this placental defect is the cause of the baby being small, but I *am* saying that there is an association between this defect and small babies.

Barker: Ideas about the immune problems of small babies come out clearly in relation to chronic obstructive lung disease. We shall hear later from Dr Martyn (1991: this volume) the evidence that lower respiratory tract infection before the age of two is a major determinant of this disease in adult life. In our Hertfordshire study, we found that poor prenatal and postnatal growth predicted mortality rates from chronic obstructive lung disease (Barker et al 1989). It would fit in with the ideas that smaller infants are immunologically compromised and therefore more vulnerable to the long-term effect of lower respiratory tract infection.

References

Barker DJP, Winter PD, Osmond C, Margetts B, Simmonds SJ 1989 Weight in infancy and death from ischaemic heart disease. Lancet 2:577–580

Björkstén B, Kjellman NI-M 1987 Perinatal factors influencing the development of allergy. Clin Rev Allergy 5:339–347

Cetin I, Marconi AM, Bozzetti P et al 1988 Umbilical amino acid concentrations in appropriate and small for gestational age infants: a biochemical difference present in utero. Am J Obstet Gynecol 158:120–156

Chandra 1975 Levels of IgG subclasses, IgA, IgM and tetanus antitoxin in paired maternal and foetal sera: findings in healthy pregnancy and placental insufficiency. In: Hemmings WA (ed) Maternofoetal transmission of immunoglobulins. Cambridge University Press, London, p 77–90

Chretien I, Pene J, Briere F, Malefijt R, Roussett F, De Vries JE 1990 Regulation of human IgE synthesis. I. Human IgE synthesis *in vitro* is determined by reciprocal antagonistic effects of interleukin 4 and interferon gamma. Eur J Immunol 20:243–251

Dutz W, Rossipal E, Ghavami H, Vessel K, Kahout E, Post G 1976 Persistent cell mediated immune deficiency following infantile stress during the first 6 months of life. Eur J Pediatr 122:117–126

Martyn CN 1991 Childhood infection and adult disease. In: The childhood environment and adult disease. Wiley, Chichester (Ciba Found Symp 156) p 93–108

Zamenhof S, van Marthens E, Margolis FL 1968 DNA (cell number) and protein in neonatal brain: alteration by maternal dietary protein restriction. Science (Wash DC) 160:322–323

Childhood infection and adult disease

C. N. Martyn

MRC Environmental Epidemiology Unit, University of Southampton, Southampton General Hospital, Southampton SO9 4XY, UK

Abstract. In England and Wales there is a strong geographical relation between current mortality from chronic bronchitis and emphysema in adults and infant mortality from bronchitis and pneumonia 50 years ago. Follow-up studies of infants and children show that certain pulmonary infections cause persisting abnormalities of lung function. This suggests that infection of an organ system during a period of rapid growth may have permanent deleterious effects. Long-term consequences of infection may also depend on age-related differences in the host response. The relationship between age of infection with hepatitis B virus and the likelihood of becoming a chronic HB_sAg carrier is an example of this. Evidence that the common communicable diseases of childhood tend to have occurred late in cases of multiple sclerosis hints at similar mechanisms in this disease. The current patterns of motor neuron disease mirror the epidemiology of poliovirus infection 40 years ago both in geographical distribution and in changes over time. The same neuronal populations are affected in both these conditions; is there a causal link?

1991 The childhood environment and adult disease. Wiley, Chichester (Ciba Foundation Symposium 156) p 93–108

The development of valvular heart disease as a sequel to rheumatic fever provides a clear example of how an infection in early life may have serious long-term effects. The purpose of this paper is to suggest that this process is far from unique. The idea that infection in infancy and childhood can initiate processes whose consequences do not become apparent until later life may be relevant to understanding the aetiology of other adult diseases.

Congenital infection

An epidemic of children with cataracts, deafness and heart defects in Australia in the 1940s first led to the recognition of the effects that infection with rubella during pregnancy had on the developing fetus. Many other features of the congenital rubella syndrome have since been described, some of which—for example, diabetes mellitus, growth deficiency and hypothyroidism—do not become apparent until childhood. Fetal damage in rubella is thought to result from infection of the placenta. Intrauterine growth is impaired and fetal metabolism deranged. The virus may also persist in the fetal ear, eye and brain.

In broad terms, the probable mechanisms are easy to understand. An agent that inhibits cell growth or causes cellular necrosis is likely to have effects that cannot later be rectified if it operates during a period when organs are developing. Successful organogenesis is critically dependent on the precise timing and sequence of a series of cell division and cell migration, each stage depending on the previous one.

Childhood respiratory infection and chronic lung disease in adult life

In England and Wales there is a strong geographical relation between current mortality from chronic lung disease in adults and infant mortality from bronchitis and pneumonia 50 years ago (see Fig. 1) (Barker & Osmond 1986). The relation is consistent in both sexes and within geographical subgroups when London and the county boroughs, urban areas and rural areas are analysed separately. The strength of the association is all the more remarkable when the known inaccuracies in death certification are taken into account.

A possible interpretation of this finding is that infection of the respiratory system in infancy has deleterious effects on the physiology of the lung that persist into adult life. Evidence in support of this view comes from follow-up studies of infants with acute lower respiratory tract infections—bronchitis, bronchiolitis or pneumonia—that have demonstrated continuing abnormalities of pulmonary function several years later (Pullan & Hey 1982, Mok & Simpson 1984). Further, in a national sample of nearly 4000 children born in 1946, the prevalence of cough during winter months at the age of 20 years was higher in those with a history of one or more lower respiratory tract infections before the age of two years than in those without such a history (Colley et al 1973).

The alternative explanation—that geographical differences in the determinants of respiratory infection at all ages have persisted unchanged over the years—is unlikely to be correct. Factors that increase the risk of respiratory infection in infancy, for example, bottle feeding, large sibship size and the presence of other children in the same room at night, cannot be operating in adults. The major known risk factor for adult respiratory disease is cigarette smoking. But mortality from lung cancer does not correlate consistently with infant respiratory mortality, which suggests that the geographical distribution of smoking differs from that of the determinants of respiratory disease in infancy and childhood.

The mechanism that leads from respiratory infection in infancy to chronic lung disease in adults is unknown. Organs may be especially vulnerable during a period of rapid growth. Minor but persisting abnormalities of pulmonary function might increase the chances of further infection, so starting a cycle in which each infection makes the next more likely.

FIG. 1.   Standardized mortality ratios for chronic bronchitis for men (*upper*) and women (*lower*) aged 35–74 during 1968–1978 and infant mortality from bronchitis and pneumonia per 1000 births in 1921–1925 in 212 areas of England and Wales. x, county boroughs; △ , London boroughs; ○, urban areas; + , rural areas. (Reproduced by kind permission of the Editor of the *British Medical Journal*.)

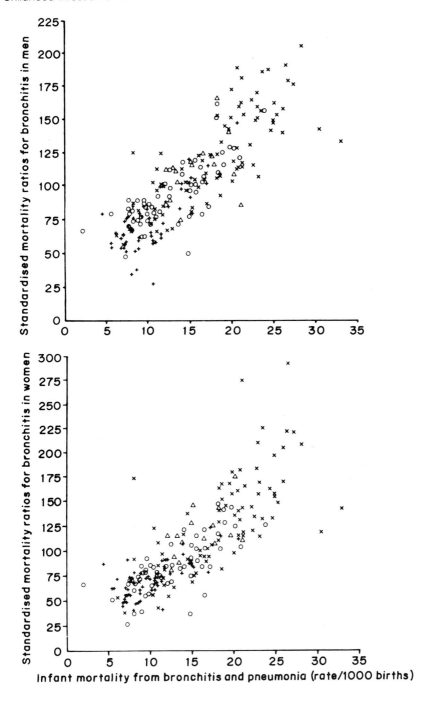

Infant mortality from bronchitis and pneumonia (rate/1000 births)

The consequences of infection may be determined by an age-dependent host response

The long-term consequences of infection may also depend on age-related differences in the response of the host. One example of this is infection with the hepatitis B virus. Infection with this virus during infancy rarely produces a hepatitic illness but commonly results in a chronic antigen carrier state (HB$_s$Ag) and an increased risk of later developing primary liver cancer. In contrast, when infection occurs in adult life hepatitis is common but the chronic carrier state rare.

It has been suggested that the aetiology of multiple sclerosis might involve an age-dependent response to a common infection (Sullivan et al 1984, Alter et al 1986, Martyn 1991). Results from several studies indicate that the age of infection with some of the common communicable diseases of childhood tends to be older in cases of multiple sclerosis than in controls (Panelius et al 1973, Compston et al 1986). Most of the reports concern measles, but there is some evidence that age at infection may also be higher with mumps and rubella.

Infection with Epstein–Barr virus in early childhood is not associated with the typical symptoms of infectious mononucleosis, whereas, if infection is delayed until late adolescence or young adult life, about 50% of cases develop the condition. In the Third World there is a high rate of seroconversion to Epstein–Barr virus before adolescence and consequently a low incidence of classical infectious mononucleosis (Warner & Carp 1981). The results of a recent case-control study of multiple sclerosis, in which an association with infectious mononucleosis was reported (Operskalski et al 1989), is therefore further evidence that common infections are acquired late in cases of the disease.

If multiple sclerosis is a rare sequel to delayed exposure to a common infectious agent, many of the unusual epidemiological features of the disease can be understood. The hypothesis can account for the rarity of the disease in the tropics, where exposure to infection in early childhood is almost invariable. It can explain why migration from areas where multiple sclerosis is common reduces the risk of developing the disease. Because of the role of the HLA system in modifying the immune response to foreign antigens, this hypothesis also has the potential to provide a mechanism for the association of the disease with particular HLA haplotypes.

Motor neuron disease as a sequel to poliovirus infection?

The possibility that motor neuron disease might be a delayed sequel to poliomyelitis has been discussed for many years. The idea is attractive because of similarities in some of the clinical features of the two diseases and because the types of neurons that are mainly affected—the first- and second-order motor neurons—are the same. The clinical picture of acute paralytic poliomyelitis is dominated by signs of lower motor neuron disease, while the majority of patients with motor neuron disease also show signs of upper motor neuron involvement. But pathological studies, both of fatal cases of poliomyelitis and of experimental poliomyelitis in primates, leave no doubt that cell loss also occurs in the upper motor neurons of the pre-central cortex and in other parts of the central nervous system (Bodian 1977).

Although some studies of series of cases of motor neuron disease have shown that the frequency of preceding poliomyelitis is higher than could be expected by chance association of the two conditions in the same individual, the argument for a causal link has always foundered because it is clear that the majority of patients with motor neuron disease cannot recall ever having suffered from paralytic poliomyelitis in the past. Nor do all cases of poliomyelitis go on to develop motor neuron disease. But such objections fail to take into account the nature of poliovirus infection. Only a small proportion of non-immune individuals exposed to the virus develop neurological symptoms and, even in those that do, the severity of the illness is very variable. It may range from a mild meningitis, through transient weakness of one limb, to generalized paralysis and death. In primates, neuronal and inflammatory lesions can be found in the nervous systems of animals that have never showed signs of infection. It is quite possible that similar subclinical forms of poliomyelitis exist in humans. Such individuals would, in all probability, never have been diagnosed as having poliomyelitis but, nevertheless, considerable loss of motor neurons would have occurred. We postulated that motor neuron disease might develop because of further neuronal loss, either through ageing or as a result of a second insult, in these subclinical cases.

To test the hypothesis we compared the geographical patterns and time trends of the two conditions (Martyn et al 1988). If there is an aetiological link, the current patterns of motor neuron disease should mirror the epidemiology of poliomyelitis half a century ago.

Several studies, both in the UK and in the USA, have shown that mortality from motor neuron disease approximates closely to its incidence (Buckley et al 1983, O'Malley et al 1987, Qizilbash & Bates 1987). Mortality can therefore be used to investigate the distribution of this disease. Poliomyelitis became a notifiable disease in 1911 and information about notification rates in individual counties and county boroughs in England and Wales was included in the Registrar General's annual reports from 1921 onwards. These data were used to explore the past patterns of poliomyelitis. The maps in Fig. 2 display the geography of motor neuron disease in the period 1968–1978 and of poliomyelitis in the period 1931–1939 for the counties of England and Wales. Table 1 shows the correlation coefficients between motor neuron disease and all notifiable infectious diseases during the same periods for the 142 local authority areas into which the counties can be divided. (These are the smallest areas for which data are available.) A positive relation between the two diseases is apparent.

98

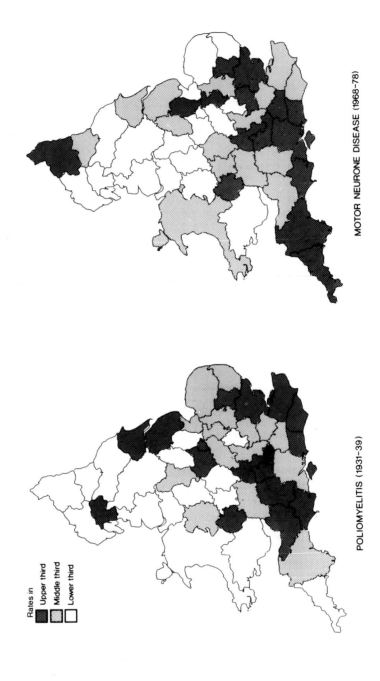

Rates in

Upper third
Middle third
Lower third

MOTOR NEURONE DISEASE (1968-78)

POLIOMYELITIS (1931-39)

FIG. 2. Standardized mortality ratios for motor neuron disease (1968–1978) and notification rates for poliomyelitis (1931–1939) in the counties of England and Wales. (Reproduced by kind permission of the Editor of *The Lancet*.)

TABLE 1 Correlations of infectious disease notification rates (1931–1939, both sexes) with mortality from motor neuron disease (1968–1978, aged 55–74, both sexes) in 142 local authority areas of England and Wales

Infectious disease	Numbers of cases notified	Correlation coefficient
Poliomyelitis	6463	0.42
Scarlet fever	946 648	−0.20
Diphtheria	506 398	−0.30
Smallpox	8548	−0.03
Enteric fever	16 981	0.11
Erysipelas	147 794	−0.18
Cerebrospinal fever	12 887	−0.08
Encephalitis lethargica	3226	0.11
Ophthalmia neonatorum	42 198	−0.02
Puerperal fever (1931–1937)	14 948	0.06

Reproduced by kind permission of the Editor of *The Lancet*.

This relation is specific: correlation coefficients between motor neuron disease and other infectious diseases are either very small or negative. Table 2 shows that this specificity is maintained when other current causes of death are correlated with rates of poliomyelitis in the 1930s.

The time trends of the two diseases also run in parallel. The epidemiology of poliomyelitis is unusual; unlike almost all other infectious diseases it responds to improvements in hygiene and social conditions by becoming commoner.

TABLE 2 Correlations of leading causes of death (1968–1978, ages 55–74, both sexes) with notifications of poliomyelitis (1931–1939, age <25 years, both sexes) in 142 areas of England and Wales

Cause of death	Correlation coefficient
All causes	−0.57
Bronchitis and emphysema	−0.50
Stroke	−0.50
Ischaemic heart disease	−0.49
Cancer of: Stomach	−0.48
Ovary	0.28
Breast	0.34
Motor neuron disease	0.42

Reproduced by kind permission of the Editor of *The Lancet*.

The explanation of this paradox can be found in the changes in the age of first exposure to the virus that accompany rising standards of living. In conditions where hygiene is poor, infection occurs during the first few months of life; at this time the infant is still partially protected by maternal antibody. The virus remains confined to the lymphoid tissue of the gastrointestinal tract because the viraemic phase necessary for the virus to reach the central nervous system is inhibited by circulating maternal antibody. As conditions improve, age of first exposure to the virus increases; the child is no longer protected by maternal antibody and the incidence of paralytic disease increases. In the population the pattern of disease changes from one of endemic infantile paralysis to recurrent epidemics of acute paralytic poliomyelitis affecting children and young adults. Over the first 50 years of this century, the UK and all other countries of the developed world experienced a striking increase in the incidence of paralytic poliomyelitis. Studies from the UK, USA and Norway (Buckley et al 1983, Martyn et al 1988, Lilienfeld et al 1989, Flaten 1989) have recently shown that motor neuron disease is now behaving in a similar way; mortality has been steadily increasing over the past two decades.

The determinants of age and likelihood of exposure to poliovirus were investigated in detail during the 1950s, when polio vaccine first became available. The information was required for the planning of programmes of immunization. As might be expected for an enterovirus whose transmission is mainly by the faecal–oral route, the major determinants are the presence or absence of running hot water, bathrooms, lavatories and other domestic amenities, and domestic crowding. We used this information to design a case-control study to test the prediction that subjects with motor neuron disease would have spent their childhood in circumstances associated with a high risk of exposure to poliovirus.

One hundred cases of motor neuron disease were compared with 335 controls matched for age and sex and selected from the lists of the same general practitioners who had first referred the cases for investigation. Information about living conditions during the first 10 years of life was obtained by a trained interviewer using a questionnaire.

The results are summarized in Table 3. Living in a house with an indoor lavatory, running hot water or a room with a fixed bath during childhood was associated with a decrease in relative risk of motor neuron disease. Living in conditions of domestic overcrowding (more than 1.5 persons per bedroom) and having three or more changes of place of residence before the age of 10 years was associated with an increase in relative risk. Factors, therefore, that decrease chances of exposure to poliovirus tend to protect against motor neuron disease, while factors that increase the chances of exposure to the virus increase the risk.

A causal relation with poliovirus infection can account for the current geographical and social patterns of motor neuron disease. It can explain the gradual increase in mortality from motor neuron disease over the past two decades. If the hypothesis is right, motor neuron disease will continue to become

TABLE 3   The relative risk of motor neuron disease according to living conditions during the first 10 years of life

	Odds ratio	*95% confidence interval*
Bathroom present	0.8	0.5–1.4
Hot water present	0.6	0.3–1.0
Indoor WC present	0.7	0.4–1.0
More than three changes of house	2.7	1.2–6.1
Crowding $\geqslant 1.5$/bedroom	1.6	0.8–2.6

commoner over the next 10–20 years until the year 2010, when people born in the 1950s, who were the first to be immunized against poliovirus, reach the age at which motor neuron disease usually presents.

Conclusions

The evidence implicating infection during infancy and childhood in the aetiology of a range of adult diseases is accumulating. Several mechanisms may be involved. The permanent damage that results from infection during a critical period of organ development is already partly understood. The idea that early infection may initiate a cycle of increasing vulnerability to further infection is more speculative. Age-related differences in host response, which may be partly related to a maturing immune system, are known to influence both short- and long-term outcome for some infections. An infection that causes loss of cells from populations that can no longer replace themselves by cell division may be important in the aetiology of motor neuron disease and, perhaps, in other degenerative diseases of the nervous system too.

References

Alter M, Zhen-Xin Z, Davanipour Z et al 1986 Multiple sclerosis and childhood infections. Neurology 36:1386–1389

Barker DJP, Osmond C 1986 Childhood respiratory infection and chronic bronchitis in England and Wales. Br Med J 293:1271–1275

Bodian D 1977 Poliomyelitis. In Minckler J (ed) Pathology of the nervous system. McGraw-Hill, New York, vol 3:2323–2343

Buckley J, Warlow C, Smith P, Hilton-Jones D, Irvine S, Tew JR 1983 Motor neurone disease in England and Wales 1959–1979. J Neurol Neurosurg Psychiatr 46:197–205

Colley JRT, Douglas JWB, Reid DD 1973 Respiratory disease in young adults: influence of early childhood lower respiratory tract illness, social class, air pollution, and smoking. Br Med J 3:195–198

Compston DAS, Vakarelis BN, Paul E, McDonald WI, Batchelor JR, Mims CA 1986 Viral infection in patients with multiple sclerosis and HLA-DR matched controls. Brain 109:325–344

Flaten TP 1989 Rising mortality from motoneuron disease. Lancet 1:1018–1019

Lilienfeld DE, Chan E, Ehland J et al 1989 Rising mortality from motoneuron disease in the USA, 1962–1984. Lancet 1:710–713

Martyn CN 1991 The epidemiology of multiple sclerosis. In: Matthews WB (ed) McAlpine's multiple sclerosis, 2nd edn. Churchill Livingstone, Edinburgh, p 3–40

Martyn CN, Barker DJP, Osmond C 1988 Motoneuron disease and past poliomyelitis in England and Wales. Lancet 1:1319–1322

Mok JYQ, Simpson H 1984 Outcome for acute bronchitis, bronchiolitis, and pneumonia in infancy. Arch Dis Child 59:306–309

O'Malley F, Dean G, Ellan M 1987 Multiple sclerosis and motor neurone disease: survival and how certified after death. J Epidemiol Community Health 41:14–17

Operskalski EA, Visscher BR, Malmgren RM, Detels R 1989 A case-control study of multiple sclerosis. Neurology 39:825–829

Panelius M, Salmi A, Halones PE, Kivolo E, Rinne UK 1973 Virus antibodies in serum specimens from patients with multiple sclerosis, from siblings and matched controls. A final report. Acta Neurol Scand 49:85–107

Pullan CR, Hey EN 1982 Wheezing, asthma and pulmonary dysfunction 10 years after infection with respiratory syncytial virus in infancy. Br Med J 284:1665–1669

Qizilbash N, Bates D 1987 Incidence of motor neurone disease in the northern region. J Epidemiol Community Health 41:18–20

Sullivan CB, Visscher BR, Detels R 1984 Multiple sclerosis and age at exposure to childhood diseases and animals: cases and their friends. Neurology 34:1144–1148

Warner HB, Carp RI 1981 Multiple sclerosis and Epstein–Barr virus. Lancet 2:1290

DISCUSSION

Thornburg: We should consider the use of correlation coefficients and the interpretation of a positive as against a negative coefficient. Some of the correlation coefficients that Dr Martyn showed were high, but negative. Do we learn something different from those values?

Barker: On its own, a correlation is simply a statistical association, positive or negative. A fundamental point is that much of epidemiology is about statistical associations which may or may not be causative. That is why epidemiological data cannot be interpreted in isolation from what is known from clinical observation and biology.

Meade: You showed both motor neuron disease and polio occurring more frequently in England and Wales in the south than in the north; but *within* the group of motor neuron disease patients, would you find the same social class distribution as is found for polio?

Martyn: We looked at social class in motor neuron disease but the picture is complicated. There is a hint of a gradual decline in the social class of people with motor neuron disease. In data for the 1960s, it seemed that the disease was commonest in social class I (Martyn et al 1988); studies made since then suggest that it is now commonest in social class III and becoming commoner in classes IV and V. Little weight should be placed on this sort of evidence, because there are no reliable data for the social class distribution of poliomyelitis.

But since polio is a disease that is largely determined by the possession or absence of domestic amenities, one might expect that it would have behaved in a similar way. That is to say, the social class of polio victims might have declined over time because it is the higher social classes who enjoy the presence of bathrooms, running hot water and other domestic amenities first.

Meade: If your hypothesis about the mechanism is correct, that there is a critical point at which one cannot sustain any further anterior horn cell loss, would you expect to find more motor neuron disease among those who have had paralytic polio, as distinct from simply seroconversion?

Martyn: If poliovirus infection were the only determinant of whether motor neuron disease develops or not, I agree that you would expect people who have had paralytic polio invariably to go on to get motor neuron disease, or at least to do so more frequently. Of course, the majority of people who have had an episode of paralytic polio do not go on to develop MND later, so a link with poliovirus can't be a complete explanation. I'm suggesting that infection with poliovirus sets up some vulnerability; that it is, perhaps, the first step in the chain of causation.

Meade: I agree that not everybody who has paralytic polio gets motor neuron disease; the question is whether *more* people with paralytic polio get motor neuron disease than people who had seroconversion as the only evidence of poliovirus infection.

Martyn: I don't know of any data that would allow an answer to that question. Case series of motor neuron disease have been looked at which show that a few per cent of cases give a history of preceding paralytic poliomyelitis. But it is difficult to estimate how often the two diseases would be associated by chance alone.

Rutter: Your findings are very interesting, but if one is building a causal hypothesis on these data, one needs to have consistent trends within populations; one way, as Tom Meade was saying, is that motor neuron disease ought to be higher in those who have had paralytic polio than after seroconversion. If that were not the case, it would cast serious doubt on the hypothesis.

If this association *is* causal, presumably one must predict a dramatic fall in motor neuron disease. When should we expect that fall?

Martyn: You are absolutely right. The final test of the idea will come about the year 2010 when the first cohort of people to have been immunized in childhood against poliovirus will reach the age at which motor neuron disease usually presents.

Hanson: Turning to multiple sclerosis, you pointed out its rarity in the tropics (p 96). One instance is the lower rate in Australia in Queensland compared to South Australia. Yet shouldn't the level of health care be consistent throughout Australia?

Martyn: That is true, and I am unable to account for the low rates of multiple sclerosis in Queensland by suggesting that these people are exposed to Third

World conditions! There is a similar situation on the West Coast of America, where the prevalence of multiple sclerosis is also low. It would be very interesting to look at the patterns of the common communicable diseases of childhood in these places.

Chandra: You correlated mortality rates from bronchitis and pneumonia in infants in the UK, 50 years ago, with current adult mortality rates from bronchitis. Clearly, the infants who died of respiratory infection are no longer in the running to develop adult respiratory disease, and it's the *morbidity* rather than *mortality* that should best correlate.

Martyn: That is right; infant mortality is being used as a marker for the incidence of infant respiratory disease.

Wadsworth: In a birth cohort of men and women born in 1946 and regularly studied from birth so far to age 43 years (Wadsworth 1990), we have found that adult chronic bronchitis and relatively poor peak expiratory flow rate at 36 years were predicted both by childhood chest illness experienced before two years, from which it was inferred that lung damage had occurred (Colley et al 1973), and by later smoking habits (Britten et al 1987)—these factors working additively and separately. By far the best predictor of childhood chest illness was poor home circumstances; even without contemporary chest illness, poor home circumstances in childhood predicted relatively low mean peak expiratory flow rates at 36 years (S.L. Mann et al, unpublished work). Our hypothesis is that experience of poor home circumstances in these years before the National Health Service (1946–1948) was associated with poor organogenesis, as shown in the poor height growth of children from these circumstances (Kuh & Wadsworth 1989), with subsequent relative high risk of illness in adult life. Risk of lower respiratory illness in early childhood was significantly increased among those living in households with other children who had chronic colds (Douglas & Blomfield 1958).

Barker: Is this effect of childhood respiratory illness and poor home circumstances stronger than the effect of smoking?

Wadsworth: Smoking was the most powerful source of attributable risk of adult chest illness, but poor home circumstances in childhood were the strongest attributable risk for low peak expiratory flow rate at 36 years (S. L. Mann et al, unpublished work).

Moxon: The proposed model relating to respiratory tract infection has potential implications of global importance. There is one prediction that could be made about infections contracted in the first two years, because the first child will often be spared exposure to infections during this time. It is the older siblings, when they go to school, who bring infections back to the younger siblings. If your hypothesis about respiratory infection is true, the relative sparing of infection in firstborn children during the first two years should be reflected in reduced respiratory morbidity in adult life. This could have important implications in considering immunoprophylaxis or other preventive strategies, the spacing of births within families, the prevention of cross-infection within families, and so on.

Barker: The picture may be complicated by measles. Data from the past suggest that the commonest antecedent of lower respiratory tract infection before the age of two was measles.

Casaer: Is there any evidence that viruses could have persisted in a long-lasting inactive or low active state in motor neurons, and that such viruses become activated by external triggers?

Martyn: Material from the CNS of cases of motor neuron disease has been examined for evidence of persisting poliovirus and to detect the presence of poliovirus-related RNA or DNA sequences. The results have been negative. It's an exciting possibility that one could investigate patients with motor neuron disease to find out whether they had been exposed to a particularly neurovirulent strain of poliovirus.

Suomi: To what extent are most of these data on infection also consistent with a genetic vulnerability hypothesis? That is, do certain individuals carry a vulnerability for the two disorders, with either no essential or an additional experiential component?

Martyn: The idea being that those who develop poliomyelitis have a gene which makes them vulnerable to poliovirus and, coincidentally, also liable to develop motor neuron disease in later life? I think that this is unlikely to be the explanation of the association that I discussed. I cannot see how a genetic hypothesis can explain the very striking trends over time.

Suomi: The question is really to what extent can a genetic hypothesis be added to the existing data set, so that there is a vulnerability, coincidental with other factors. Is it possible that in some cases you may have an age-dependent disorder that will be expressed only in certain individuals who carry this genetic vulnerability?

Martyn: If you are saying that these conditions are likely to have a genetic component, then I agree; it would be surprising if there weren't genetic as well as environmental determinants of the consequences of exposure to micro-organisms.

Wadsworth: In the 1946 birth cohort study the predictive power of parental and grandparental bronchitis for bronchitis in the study member is good (S. L. Mann et al, unpublished work), but we believe this is probably environmental rather than genetic, and we expect the relationship to lessen in later generations as the prevalence of smoking is reduced. We hope in due course to be able to examine genetic risk from blood samples, and then we shall be able to compare the predictive value of lifetime environmental data with the genetic information.

Wood: Chris Martyn's message was underlined for me earlier this morning, because before the meeting I was looking at a six-week-old Turkish refugee baby who was failing to thrive and now has very severe pneumonia and is on a ventilator. This is a striking example of a baby experiencing a lot of lung damage very early. The X-rays are very abnormal and I have no doubt that, over the next few years, this baby will be symptomatic with wheezing.

More difficult is the fact that paediatricians, myself included, tend to trivialize the much more common minor respiratory infections of childhood; a very significant part of the work of a general practitioner in the UK may be in respiratory infections of childhood. I am therefore looking for a message: shouldn't we be trying to have a much more active preventive concept, at the social, microbiological and immunological levels? Shouldn't we be trying harder to prevent these apparently minor illnesses? The only reason for *not* doing so would be the possibility of postponing bronchiolitis, say, to the age of 25 years, like the time shift in the peak incidence of poliomyelitis in developed communities before immunization was established.

Richards: Small premature babies have a lot of respiratory problems; is this just more bad news for them?

Wood: Very low birth weight babies are a very small proportion of the total UK child population, still. Their wheezing may get better by five or six years of age, but that's as far as we can say, as yet. Whether respiratory viruses significantly infect preterm babies in neonatal units is a matter of current concern and interest.

Murray: Presumably, when young women, potential mothers, enter a Western society as immigrants from underdeveloped countries, they may be exposed to viruses that they may not have experienced before. As I understand it, congenital rubella infection was very common in the children of Afro-Caribbean immigrants to Britain in the 1960s. Is there evidence of the children of other immigrant groups having more congenital disorders as a consequence of acquiring infections *in utero* to which their mothers had not previously been exposed?

Martyn: I do not know; it is an interesting possibility.

Golding: Data from the 1970 birth cohort in Great Britain indicate that the children who get early chest illness have all sorts of ways in which they could be expected to be vulnerable: for example, they are more likely to have had a delay in the onset of regular respirations, and they are more likely to have been growth retarded, and to be delivered preterm. It is therefore dangerous to say that the chest illness is the 'cause' of the adult chronic bronchitis. It could be a vulnerability factor running through the whole story.

On the geographic variation in the incidence of polio and motor neuron disease, the association seems to be mainly based on an increased incidence in the south of England, which is an unusual geographic pattern. Indeed, the geographic patterns for chicken pox, as well as allergic disorders such as eczema, in Great Britain in 1980 were very similar (unpublished observation). If we had incidence data on other childhood infections we might find other similar patterns.

I was wondering whether the incidence pattern of appendicitis had been examined, because that story sounded very much the same as that of polio, with a sudden increase with increasing standards of living. Does it show the same geographic pattern as motor neuron disease?

Barker: The basic idea behind our work on appendicitis is that improvements in Western hygiene show three phases. The first phase is that of 'Third World' hygiene, where high levels of exposure to infection occur very early in life. Then follows an intermediate phase when conditions have improved. Exposure to infection is delayed past infancy, when there is protection by maternal antibodies, and occurs in early childhood. That is the stage where paralytic polio is seen. The third phase is when living conditions have improved further and the total dose of exposure to infection at any age is greatly reduced.

This is an interesting model, because it can explain why a disease which is an age-dependent consequence of infection could rise and fall in its incidence (Barker 1989). The mystery about appendicitis is that wherever it appears, it then disappears. So there is a seductive analogy with these ideas about polio (Barker 1985), and it's a line of research in which you can show rather strong predictions in individuals (Coggon et al 1991); Michael Wadsworth and Jean Golding have been associated with this work (Barker et al 1988). The past geography of appendicitis in the UK is, however, obscure, so we can't address the particular point raised by Jean Golding.

Chandra: Coming back to the relationship between infant mortality and adult bronchitis, the correlations are just that—correlations, rather than a cause-and-effect relation. However, the possibility of a common cause for both conditions has to be considered. There are several rare but well-known conditions which could cause both these types of illnesses in children and in adults, such as α_1-antitrypsin deficiency (about 1:1000 births), or mild forms of immunodeficiency (about 1% of the population). A common underlying factor could explain childhood illness and subsequent respiratory problems in adults. Finally, Dr Pryse Phillips suggested a correlation between canine distemper virus and multiple sclerosis. How does that fit into the geographic and time frames that you have looked at?

Martyn: An alternative interpretation of the correlations is that the determinants of respiratory infection in childhood persist and cause respiratory infection in adult life. This is unlikely to be a very large part of the explanation, because, as you say, your examples are rare ones and account for only a small proportion of respiratory disease. Also, we know that the major determinants of respiratory disease in childhood are *not* the same as the major risk factors for respiratory disease in adult life.

The canine distemper story began with a remarkable report of three sisters whose first symptoms of multiple sclerosis occurred within a very short time of their pet dog developing an encephalitic illness. The idea has been pursued extensively in case-control studies. The balance of the evidence is overwhelmingly against exposure to dogs being important in multiple sclerosis. And the geographic distribution of the disease suggests that a simple explanation, such as close contact with domestic animals, cannot be correct.

Chandra: A map of the world prevalence of these two diseases shows a strong correlation.

Martyn: The absence of an association between exposure to dogs and risk of multiple sclerosis when tested in individuals rather than in populations is a decisive refutation of the hypothesis.

References

Barker DJP 1985 Acute appendicitis and dietary fibre: an alternative hypothesis. Br Med J 290:1125–1127

Barker DJP 1989 Rise and fall of Western diseases. Nature (Lond) 338:371–372

Barker DJP, Osmond C, Golding J, Wadsworth MEJ 1988 Acute appendicitis and bathrooms in three samples of British children. Br Med J 296:956–958

Britten N, Davies JMC, Colley JRT 1987 Early respiratory experience and subsequent cough and peak expiratory flow rate in 36 year old men and women. Br Med J 294:1317–1320

Coggon D, Barker DJP, Cruddas M, Oliver RHP 1991 Housing and appendicitis in Anglesey. J Epidemiol Community Health, in press

Colley JRT, Douglas JWB, Reid DD 1973 Respiratory disease in young adults: influence of early childhood lower respiratory tract illness, social class, air pollution, and smoking. Br Med J 2:195–198

Douglas JWB, Blomfield JM 1958 Children under five. Allen & Unwin, London

Kuh DJL, Wadsworth MEJ 1989 Parental height, childhood environment and subsequent adult height in a national birth cohort. Int J Epidemiol 3:663–668

Martyn CN, Barker DJP, Osmond C 1988 Motoneuron disease and past poliomyelitis in England and Wales. Lancet 1:1319–1322

Wadsworth MEJ 1990 The imprint of time. Oxford University Press, Oxford

Critical periods in brain development

James L. Smart

Department of Child Health, University of Manchester, The Medical School, Oxford Road, Manchester M13 9PT, UK

Abstract. The growth of the brain after organogenesis can be described as occurring in two somewhat overlapping phases: a phase of neuronal multiplication followed by one of glial proliferation, during and after which occur myelination and dendritic and axonal arborization. Within this gross chronology is a finer-grained chronology, with, for instance, different neuronal populations dividing at different times. The course of brain development can be affected by a variety of factors, the nature and extent of the perturbation dependent on the timing of the treatment with respect to stage of brain development. Growth processes completed before treatment are unaffected. Only those processes occurring at the time of the treatment are affected, plus some later-occurring processes, as a result of a cascade of effects. These concepts are examined briefly with reference to ionizing radiation, hormones and environmental stimulation and more fully with respect to nutrition. Undernutrition appears to depress the rate of all brain growth processes contemporaneous with it to the same extent. Whether the effects produced are likely to be permanent is discussed, together with the possibility that there may be mechanisms that attenuate or compensate for adverse effects.

1991 The childhood environment and adult disease. Wiley, Chichester (Ciba Foundation Symposium 156) p 109–128

Critical, sensitive and vulnerable periods

The terms 'critical' and 'sensitive' period are often used interchangeably, though their meaning and implications are rather different. In fact, these terms are very rarely defined and the reader is usually left to derive their meaning from accounts of experiments which exemplify their characteristics. 'Sensitive period' is preferable, since its non-scientific meaning accords better with the meaning implied in its scientific usage than is the case with 'critical period'. The term 'sensitive period' will be adopted here, defined as 'a stage at which some aspect of development occurs with greatest ease'. The term 'critical period' has connotations of crisis and abruptness, which are appropriate only in certain instances. It is best avoided or used only as a special class of sensitive period.

109

The term 'vulnerable period', coined by Davison & Dobbing (1966), focuses on harmful consequences and ignores the converse possibility that the period is likely also to be a time of opportunity to realize a full flowering of developmental potential. 'Sensitive period' avoids this imbalance.

There is evidence going back to ancient times of an awareness of at least one phenomenon with a sensitive period. This is *imprinting*: 'the learning process involved in developing . . . the tendency to follow or otherwise approach an object (usually another animal and in ordinary development one of the animal's own species)' (Barnett 1963). Apparently Pliny tells of 'a goose which followed Lacydes as faithfully as a dog' and Reginald of Durham in the twelfth century records eider ducks following human beings (cited in Thorpe 1963). Whether these authors also knew that there was a sensitive period for the development of the 'following responses' is not recorded. The first systematic observations of imprinting, including evidence of its age dependence, were published by Spalding (1873). Chicks hatched in the absence of a hen were found to follow any moving object to which they were exposed in the first few days after hatching, but not later.

The first evidence of sensitive periods, therefore, derived from ethology. The second line of evidence came some decades later from quite a different quarter, from a branch of embryological study which has come to be known as teratology. Investigations of factors causing cyclopia in fish (Stockard 1921) and inhibited head development in planaria (flatworms) (Child 1921) both indicated the existence of sensitive periods. Indeed, Child (p 32) felt able to conclude that 'in all cases the differences in susceptibility correspond to differences in structure, rate of growth, development and differentiation'. I suppose that this was a very general 'general conclusion'. Nevertheless, it is humbling to note that the plethora of research since then has left it virtually unscathed.

Recently, Bornstein (1989) suggested a framework for research and theory concerning sensitive periods, which relates principally to the development of behaviour but may also be applicable to structural development, including brain development. His framework comprises four parameters (timing, mechanisms, consequences, and evolutionary and ontogenetic time scales) with a total of 14 characteristics. The advantage of this systematic approach is that it quickly reveals the gaps in our knowledge and suggests avenues for research.

Chronology of brain growth and development

The sequence of developmental processes in the brain seems to be the same across mammalian species. Some of these processes follow one after the other, others overlap and yet others occur in parallel. A gross simplification is that the quantitative bulk of neuronal proliferation precedes that of glial proliferation which in turn precedes myelination, dendritic branching and synaptogenesis, but this obscures much important detail. Figure 1 shows the timing of appearance

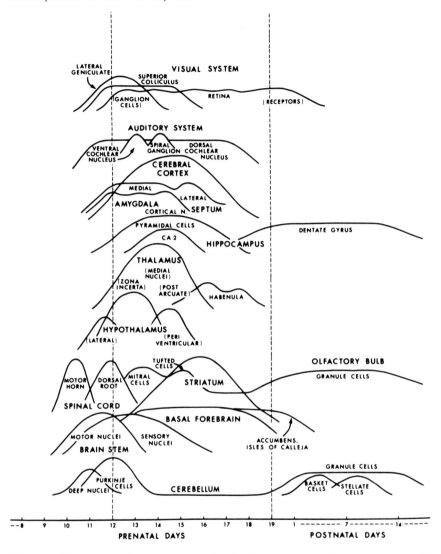

FIG. 1. Chronology of neuron production in the mouse central nervous system. The
vertical line on prenatal Day 12 represents the last day of gestation when gross external
malformations can be induced by interference (non-nutritional) with cell proliferation.
The vertical line on Day 19 represents the day of birth. (Reproduced with kind permission
of the publisher and author, from Rodier 1980.)

of the various neuronal cell populations in mouse brain (Rodier 1980). In most
regions this is prenatal, the only exceptions being cerebellum, hippocampus and
olfactory bulbs, in which microneuronal proliferation occurs postnatally.
Myelination too occurs at different times in different parts of the brain. For

instance, in man, myelination of the stato-acoustic tectum and tegmentum is complete before birth, whereas that of the reticular formation and great cerebral commissures continues till at least seven years of age (Yakovlev & Lecours 1967).

Always, the period of fast brain growth occurs early in life, in advance of the growth spurt for the body as a whole.

The concept of timing

Given that the notion of 'timing' is fundamental to the concept of sensitive periods, it is pertinent to enquire what we mean by 'timing'. Timing with respect to what? To age from birth or even conception, or to stage of development? In actual fact, we may mean any of these, depending upon which developing process we are concerned with. For instance, the sensitive periods for imprinting of chick on mother hen or lamb on ewe are best related to time of hatching and of birth respectively. In contrast, the sensitive periods for brain growth processes are best related to the stage of development, for the simple reason that the timing of brain growth with respect to birth varies greatly between species. In precocial species like the guinea pig, most of brain development occurs prenatally, whereas altricial species which are relatively immature at birth, like rat, mouse and man, are largely postnatal brain developers. It certainly behoves us, in talking about timing, to choose a frame of reference which is appropriate to the developmental process under consideration.

Most of the remainder of this review is confined to the subject of sensitive periods in brain growth, so the timing of any treatment will be expressed relative to the stage of brain development.

Sensitive periods in brain growth

General principles

I propose a number of general principles below, all of which are straightforward and one or two of which may even be thought self-evident. They are, nevertheless, worth stating. They assume that any 'treatment' encountered during development is not so catastrophically severe as to cause macroscopic lesions (not anoxia or physical damage, for example). The principles are as follows:

(i) Those growth processes completed before the treatment is applied will be unaffected.

(ii) Only those processes occurring at the time of the treatment will be affected, plus some later-occurring processes, as a result of a cascade of effects.

(iii) As a corollary of (ii), only periods in which some brain growth normally occurs can be sensitive periods for brain growth.

(iv) The extent of the effect on brain growth will depend on the intensity (severity) of the treatment and the amount of growth that would normally have occurred during the period of treatment.

These principles are examined below in relation to a number of possible influences, and most fully with respect to undernutrition. A fifth principle is proposed in Section (iii) below.

Examples of factors with effects that depend on when they are applied

(i) Agents which kill dividing cells or arrest cell division. By and large, these are not factors which the animal will encounter during development under anything like normal circumstances. They are factors such as ionizing radiation, which kills dividing cells, and a number of toxins that interfere with the process of mitosis. Nevertheless, the study of their effects in certain human pathological situations and in experiments with laboratory animals is instructive. For instance, epidemiological evidence on the occurrence of mental retardation in the survivors of the atomic bomb attacks on Japan in 1945 suggests a sensitive period in brain development. The incidence of mental retardation corresponds

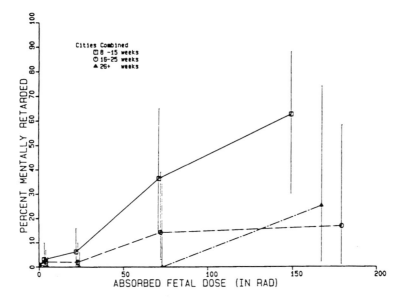

FIG. 2. Frequency of mental retardation among those exposed *in utero* to ionizing radiation from atomic bombs in Japan, by radiation dose and gestational age. Data for Hiroshima and Nagasaki combined. Vertical lines indicate 90% confidence intervals. There was no mental retardation among those exposed between 0 and 7 weeks, so that group is not shown. (Reproduced with kind permission of the publisher and authors, from Otake & Schull 1984.)

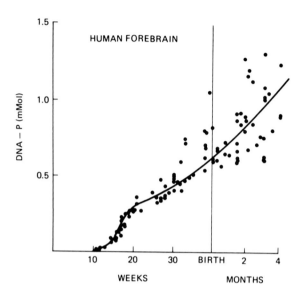

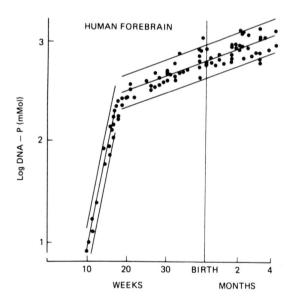

FIG. 3.   Total DNA-phosphorus (cell number) in human forebrain from 10 gestational weeks to four postnatal months, showing the two-phase characteristics of prenatal cell multiplication. The lower graph is a semi-logarithmic plot of the data in the upper graph, with regression lines and 95% confidence limits added. (Reproduced with kind permission of the publisher and authors, from Dobbing & Sands 1973.)

both to the stage of development of fetuses at the time they were exposed and to radiation dose (Otake & Schull 1984). The greatest incidence is highly correlated with exposure between eight and 15 weeks of gestation (Fig. 2). There were no cases associated with earlier exposure and very few with exposure later than that. In fact, the sensitive period corresponds closely with the probable period of neuroblast multiplication (Fig. 3; Dobbing & Sands 1973). It seems likely that the radiation killed dividing neuroblasts, thus reducing the extent of the dividing cell population and hence restricting the final number of neurons.

Results from animal experiments strongly support the above conclusions. For example, rats exposed to an inhibitor of cell division (hydroxyurea) at the beginning of the major period of brain neurogenesis (Day 14 of gestation) show a 30% deficit in brain weight both at birth and in adulthood and perform poorly in tests of learning as adults (Adlard & Dobbing 1975). Interestingly, treatment with another inhibitor prenatally (gestation Day 14) has a relatively greater effect on forebrain than on cerebellum growth, whereas the converse is true of postnatal treatment (postnatal Day 5) (Adlard et al 1975). This corresponds nicely with the different timing of neurogenesis in the two regions (Fig. 1).

Another factor whose effects resemble those of these noxious physical and chemical agents, and which occurs naturally, is hyperthermia. Raised body temperature in pregnant guinea pigs can stunt fetal brain growth, apparently through effects on mitotic cells, and impair subsequent discrimination learning, provided that the heat stress is correctly timed to coincide with neuroblast multiplication (Edwards et al 1976).

(ii) Hormones. During the late 1960s and throughout the 1970s a great deal of research was conducted on the effects of thyroid hormones and corticosteroids on brain development. Much of this is summarized conveniently in Table 1, taken from Balazs (1973) (for other reviews see Patel et al 1980, Legrand 1980). It should be pointed out that all of these studies were on rats and mice, in which the brain growth spurt is postnatal, and aimed to examine the effects of hormone treatment on the processes making up this phase of brain development. In most cases, therefore, the treatment was postnatal or was started shortly before birth. These studies were not specifically designed to investigate sensitive periods.

Corticosteroids are found to slow the rate of cell acquisition during and for a short time after the period of treatment, and, because there is no evident extension of the period of cell formation, final cell number remains deficient. The effects of thyroid hormones are more interesting, because they are found to influence the timing of the offset of the period of cell proliferation. In general, thyroid deficiency retards development, whereas treatment with thyroid hormone accelerates it (Table 1); but so too do they respectively retard and advance the age of offset of the period of cell formation, such that in deficient animals the final cell number attains normality through prolongation of cell division, in contrast to thryoid hormone-treated animals in which final cell number is

TABLE 1 **Effects of thyroid hormone and corticosteroids on postnatal brain development in rats and mice, as summarized by Balazs (1973)**

	Thyroid deficiency	*Treatment with thyroid hormone*	*Treatment with corticosteroids*
Postnatal cell formation			
Final cell number	Normal	Deficit	Deficit
Rate of cell acquisition	Retarded	No effect in first week	Retarded in first week
Period of extensive cell formation	Prolonged (cerebellum)	Premature termination	Normal
Biochemical maturation			
Conversion of glucose carbon into amino acids	Retarded	Advanced	Normal
Development of metabolic compartmentation	Retarded	Advanced	—
Neuronal differentiation			
Dendritic branching and spine formation	Retarded	Accelerated	Transiently retarded
Synaptogenesis	Retarded	Accelerated (but final number decreased)	—
Behavioural			
Appearance of innate behaviour patterns	Retarded	Advanced	Retarded
Performance in tests of adaptive behaviour	Impaired	Impaired	Normal

Reproduced by kind permission of the author and publisher. See Balazs (1973) for references.

deficient because of the premature termination of cell division. As a result of the altered timing of the various developmental processes with respect to one another, however, the 'normal' final cell number in the deficient animals is made up of abnormal relative numbers of the different cell types. Patel et al (1980) suggest that the effects of thyroid hormone on brain cell acquisition are brought about through changes in cell migration and/or cell differentiation rather than alterations in the generative cycle of dividing cells.

Gonadal hormones too appear to exert an influence on brain development, though in a more subtle way than the hormones considered above. It has been known for some time that there are well-defined sensitive periods for the development of sexual dimorphism in behaviour in many species of mammal and that these are sex hormone (most notably androgen) dependent. For instance, castration of male rat pups within the first few days after birth, but not later, has a lasting feminizing effect, whereas administration of androgen to female pups at the same stage has a masculinizing effect (reviewed by Bermant & Davidson 1974). This relates particularly to sexual behaviour, but also to other aspects of behaviour which normally differ between the sexes (reviewed by Goy

& McEwen 1980). The most effective period for modification varies between species and does not bear a constant relationship with birth, being prior to birth for the guinea pig and rhesus monkey, both before and after birth for the dog, and shortly after birth for the rat, mouse, hamster and ferret (Goy & McEwen 1980).

Until the early 1970s the existence of a morphological basis in the brain for sexually dimorphic behaviour was thought to be unlikely. Since then, evidence has accumulated of what is probably best described as 'differential neuronal growth' in the preoptic area and hypothalamus. Differences in type of synapse (Raisman & Field 1973), shape of dendritic tree, and size of nucleus (group of nerve cells) seem to be involved (reviewed by Goy & McEwen 1980).

Sensitive periods for the development of sexual dimorphism in behaviour seem to be among the few instances when it is correct to use the term 'critical period', because they are clearly delineated in time and have dramatic effects for the life of the animal.

(iii) Undernutrition.   One point which is worth making, both because it is of practical significance and for comparative purposes, is that the fetus *in utero* is to a large extent buffered by its mother from the effects of undernutrition. This is in contrast to the effects of ionizing radiation, for instance, against which the mother affords virtually no protection. Hence it is difficult to affect the prenatal stages of brain development by undernutrition, unless it is very severe and chronic, perhaps commencing some considerable time before conception. The developing animal is much more liable to be undernourished after birth, and the remainder of this section is devoted to a consideration of the effects of postnatal undernutrition. All of the research recounted is on the rat, which is a largely postnatal brain developer and is rather less mature in terms of brain development at birth than the human baby (Dobbing & Sands 1979).

There is plenty of evidence of restriction of various aspects of brain growth during undernutrition (Bedi 1984, Dodge et al 1975, Wiggins 1982). What is more interesting is to enquire whether there are synthetic processes in the brain which are more sensitive (vulnerable) to undernutrition than others. In a sense, this would be the converse of the well known 'brain sparing' phenomenon (Stewart 1918) whereby the growth of the developing brain is less retarded by undernutrition than that of the rest of the body in spite, presumably, of being bathed in the same fluids carrying the same nutrients. The question then is whether there are developmental processes within the brain that are especially sensitive to undernutrition. Most would answer that, of course, there are, and go on to cite the 'selective' effect on the cerebellum. Let us examine this example.

There does indeed appear to be a greater effect of undernutrition during the suckling period in the rat on cerebellum than on the rest of the brain. For instance, such undernutrition can result in a deficit in total DNA (an index of cell number) of 10% in forebrain but 15% in cerebellum (Lynch 1973). It is illuminating to examine the curves for the accumulation of DNA in forebrain

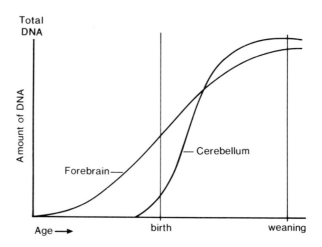

FIG. 4. Stylized curves showing the accumulation of DNA in rat forebrain and cerebellum between conception and weaning. (Adapted from Lynch 1973.)

and cerebellum of the rat (Fig. 4). About half of the adult amount has been attained by birth in the forebrain, but only 3% in the cerebellum (Balazs et al 1971). That is, in our example the forebrain has attained half its complement of DNA before the onset of undernutrition (principle (i), p 112). Let us propose further that the undernutrition commencing at birth continues till the end of brain growth and that it causes an arbitrary 15% growth restriction that is uniform throughout the brain. The 15% restriction in the remaining 50% of forebrain DNA accumulation will result in an overall deficit in forebrain DNA of 7.5%. A similar calculation for the cerebellum yields an overall deficit of 14.55%. The cerebellum is only 'vulnerable' because nearly all of its growth falls within the period of undernutrition and not because its growth is especially sensitive (exemplifying principle (iv), p 113). Careful examination of the available evidence led to the conclusion that this simple relationship was general and to the proposal that *undernutrition affects the rate of all synthetic processes in brain contemporaneous with it to the same extent* (Peeling 1982, Peeling & Smart, in preparation).

This generalization has the advantages of simplicity, explanatory power and testability, and hence seems worthy of consideration. It remains to be seen how well it will stand up to scrutiny; so far it has withstood the challenge of certain apparent exceptions which have proved to be explicable (Smart 1990, Peeling & Smart, in preparation).

So far, I have discussed effects occurring at the time of the undernutrition. It is important to go on to enquire what is the extent of recovery after refeeding ('recovery' being defined as 'diminution of deficit in absolute terms'). One question that bears on this is whether development can be delayed or prolonged in response to undernutrition, such that a normal amount of development might be achieved but over a longer period. What little evidence there is on this point appears to be contradictory, with the timing of cell acquisition largely immutable (Dobbing & Sands 1971) but that of synapse formation extendable far beyond the usual period (Warren & Bedi 1984). Whether or not delay is possible may well differ between one growth process and another. It proves to be very difficult to propose a useful general hypothesis on recovery from undernutrition because of the complex nature of brain development, with different processes sometimes dependent upon one another and to a greater or lesser extent at different stages. Just as the theoretical approach is problematical, so too is it extremely difficult to derive a general conclusion from the available evidence (Peeling 1982, Peeling & Smart, in preparation). The best approach would appear to be to treat each particular process or system individually, and to make predictions based on a detailed knowledge of brain development.

An instructive example, nicely exemplifying principles (i) and (ii) (p 112), is that of connectivity between neurons, as indicated by the extent and pattern of dendritic branching. The orderly anatomical arrangement of the cerebellum makes it an especially good site for such study. The only output from the cerebellum is from the large Purkinje cells (large neurons that are relatively few in number), which receive inputs from many granule cells (microneurons) and from the olivary nucleus. Purkinje cells arise wholly prenatally in the rat and the micro-neurons of the cerebellum very largely postnatally (Rodier 1980). One would predict, therefore, that undernutrition during the suckling period would have no effect on the numbers of Purkinje cells but a substantial and probably lasting effect on the numbers of microneurons. This would appear to be borne out by the published evidence, though it has to be admitted that this is circumstantial and not absolutely conclusive (Bedi 1984). Certainly, persisting deficits have been found in the numbers of granule cells per section examined (McConnell & Berry 1978a). However, it is more reliable to use measures of 'numerical density' of structures and ratios of one structure to another (Bedi 1984).

The density of granule cells in cerebellar cortex differs little between control and undernourished rats either at the end of undernutrition or after a lengthy period of refeeding, whereas the density of Purkinje cells is much greater in undernourished rats at both stages (Table 2), presumably because their cerebella have much the same number of Purkinje cells but are filled out with less other material. The resulting ratio of granule cells to Purkinje cells is permanently reduced in previously undernourished rats by some 25% (Bedi et al 1980a). This might have implications for the development of the Purkinje cell dendritic tree, and this indeed proves to be the case. There is evidence of some dendritic remodelling during rehabilitation from undernutrition, but there remains a deficit of 28% in the extent of Purkinje cell dendritic networks, due primarily to a deficit in segment number (McConnell & Berry 1978b). An important part of the explanation for this derives from the likely relationship between the

TABLE 2   Numerical densities of cerebellar granule and Purkinje cells and granule-to-Purkinje cell ratios in control and experimental rats

	Age of rats (days)	Control	Experimental[a]	% difference
Numerical density of granule cells in whole cortex ($\times 10^3$) per mm^3	30	1187 ± 88	1122 ± 70	-5
	160	768 ± 37	780 ± 62	$+2$
Numerical density of Purkinje cells (per mm^3)	30	3082 ± 263	3960 ± 211	$+27*$
	160	2329 ± 112	3185 ± 131	$+37*$
Granule-to-Purkinje cell ratios	30	395 ± 34	290 ± 27	$-27*$
	160	335 ± 28	250 ± 23	$-25*$

Results are mean $\pm$ SE. Data from Bedi et al (1980a).
[a]Experimental rats were undernourished from birth to 30 days of age, after which some were nutritionally rehabilitated until 160 days of age.
*$P < 0.05$, Student's t-test.

number of segments in a Purkinje cell dendritic network and the local density of growing axons, which is a function of granule cell number.

In summary, the principal effect of undernutrition is to slow down the rate of concurrent brain growth processes, causing deficits in growth attainment. Distortion of brain growth can result from undernutrition which is not uniform in severity over the whole period of brain growth or which covers only a part of it.

(iv) The environment.   The influence of the environment will be mentioned only briefly here, since it is dealt with extensively in relation to the visual system in another chapter (this volume: Blakemore 1991). It certainly seems to be the case that environmental stimulation specific to a particular sensory modality exerts an important influence on the anatomical development of appropriate parts of the brain during sensitive periods early in life. What is perhaps more surprising is that giving animals a more generally 'enriched' environment after the period of fast brain growth has demonstrable effects on a variety of brain measures, especially in occipital cortex (Rosenzweig & Bennett 1978). Thus, rats reared after weaning in social groups in physically complex environments show greater weight of cerebral cortex and depth of occipital cortex, and enhancement of various aspects of dendritic branching, than controls reared singly in small, featureless cages (Bedi & Bhide 1988). While these effects are typically small, the weight of evidence for their existence is compelling. Also, they are demonstrable even after later periods of enrichment, and hence are not confined to a sensitive period. We can perhaps take comfort that, in these respects, the brain seems to remain sensitive well into adulthood.

Possible attenuating or compensatory mechanisms

Viewed from a biological standpoint, it would not be surprising if animals had evolved mechanisms to attenuate or compensate for the adverse effects of restricting or damaging environmental conditions during development. This is considered briefly below in relation to undernutrition (discussed at greater length by Smart 1990).

Delayed development

One obvious possibility is that there could be an advantage to the undernourished animal in prolonging development, such that the same amount of growth accrued but over a longer period, or of postponing development till nutrition was more favourable. These mechanisms would allow the undernourished animal to attain the same end-point of development as its well-fed counterpart. It would appear, however, that such mechanisms operate to only a limited degree with respect to gross indices of brain growth. Little shift is found in the timing of the peak velocities of accretion of whole brain weight, DNA or cholesterol in rats growth-restricted from birth (Fig. 5), and no catch-up growth in these measures on refeeding (Dobbing & Sands 1971). For such characteristics, the brain seems

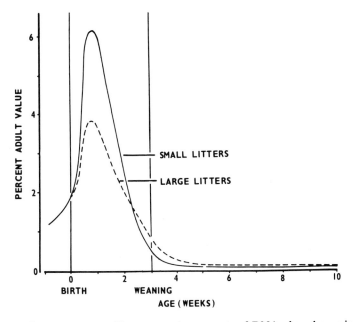

FIG. 5.   Relative rate curves of increments in amounts of DNA-phosphorus in whole brain of rats reared in small (well-fed) or large (underfed) litters. (Reproduced with kind permission of the publisher and authors, from Dobbing & Sands 1971.)

to have a once-only opportunity for development. The same may not be true
of the brain's finer structure, in that considerable catch-up is found in the
numbers of synapses per neuron in the visual cortex of rats refed after prolonged
undernutrition (Warren & Bedi 1984).

Altered cell cycle characteristics

A mechanism attenuating the effects of undernutrition on cell proliferation in
the brain was discovered by Lewis et al (1975). They found, perplexingly, that
rates of cell production in the germinal layers were only slightly slowed by
undernutrition, in spite of the DNA-synthetic (S) phase in the dividing cells being
markedly prolonged. The answer to the paradox was that the G1 phase of the
cell cycle (a non-synthetic phase) was greatly shortened in undernourished rats,
thus largely compensating for their longer S phase and maintaining cell
production at a more normal rate than would otherwise have been the case.

Reduced loss of neurons and synapses

It is now well known that there is an exuberant overproduction of both neurons
and synapses in some parts of the brain during development and that a
proportion of these are lost subsequently (Herschkowitz 1988). Supposing that
nutritional growth restriction results in deficits in the numbers of neurons and
synapses produced, the extent of the deficits could be diminished if loss of these
structures was relatively less in undernourished than in well-fed animals. There
is certainly clear evidence of this with respect to synapses. The synapse-to-neuron
ratio in frontal cortex of previously undernourished rats shows considerable
recovery, apparently through this mechanism (Table 3; Bedi et al 1980b). This
sparing of synapses seems potentially a most important phenomenon. How
general it is within the brain, whether it would occur during continued
undernutrition, and whether it could be promoted, for instance, by providing
a stimulating environment, are all questions deserving further investigation.

TABLE 3   Synapse-to-neuron ratios ($\times 10^3$) in frontal
cortex of well-fed rats and rats undernourished from birth
to 30 days

Age (days)	Control	Experimental[a]	% difference
30	22.3 ± 3.3	14.0 ± 1.5	-37*
160	13.4 ± 1.1	11.8 ± 0.7	-12

Results are mean $\pm$ SE. Data from Bedi et al (1980b).
[a]Experimental rats were undernourished from birth to 30 days
of age, after which some were nutritionally rehabilitated until
160 days of age.
*$P < 0.05$, Student's t-test.

Acknowledgements

I am grateful to the Medical Research Council and to the National Fund for Research into Crippling Diseases for financial support.

References

Adlard BPF, Dobbing J 1975 Maze learning by adult rats after inhibition of neuronal multiplication *in utero*. Pediatr Res 9:139–142

Adlard BPF, Dobbing J, Sands J 1975 A comparison of the effects of cytosine arabinoside and adenine arabinoside on some aspects of brain growth and development in the rat. Br J Pharmacol 54:33–39

Balazs R 1973 Hormonal influences on brain development. Biochem Soc Spec Publ 1:39–57

Balazs R, Kovacs S, Cocks WA, Johnson AL, Eayrs JT 1971 Effect of thyroid hormone on the biochemical maturation of rat brain: postnatal cell formation. Brain Res 25:555–570

Barnett SA 1963 A study in behaviour. Methuen, London

Bedi KS 1984 The effects of undernutrition on brain morphology: a critical review of methods and results. Curr Top Res Synapses 2:93–163

Bedi KS, Bhide PG 1988 Effects of environmental diversity on brain morphology. Early Hum Dev 17:107–143

Bedi KS, Hall R, Davies CA, Dobbing J 1980a A stereological analysis of the cerebellar granule and Purkinje cells of 30-day-old and adult rats undernourished during early postnatal life. J Comp Neurol 193:863–870

Bedi KS, Thomas YM, Davies CA, Dobbing J 1980b Synapse-to-neuron ratios of the frontal and cerebellar cortex of 30-day-old and adult rats undernourished during early postnatal life. J Comp Neurol 193:49–56

Bermant G, Davidson JM 1974 Biological bases of sexual behavior. Harper & Row, New York

Blakemore C 1991 Sensitive and vulnerable periods in the development of the visual system. In: The childhood environment and adult disease. Wiley, Chichester (Ciba Found Symp 156) p 129–154

Bornstein MH 1989 Sensitive periods in development: structural characteristics and causal interpretations. Psychol Bull 105:179–197

Child CM 1921 The origin and development of the nervous system. University of Chicago Press, Chicago

Davison AN, Dobbing J 1966 Myelination as a vulnerable period in brain development. Br Med Bull 22:40–44

Dobbing J, Sands J 1971 Vulnerability of developing brain. IX. The effect of nutritional growth retardation on the timing of the brain growth-spurt. Biol Neonate 19:363–378

Dobbing J, Sands J 1973 The quantitative growth and development of the human brain. Arch Dis Child 48:757–767

Dobbing J, Sands J 1979 Comparative aspects of the brain growth spurt. Early Hum Dev 3:79–83

Dodge PR, Prensky AL, Feigin RD 1975 Nutrition and the developing nervous system. Mosby, St Louis, Missouri

Edwards MJ, Wanner RA, Mulley RC 1976 Growth and development of the brain in normal and heat-treated guinea pigs. Neuropathol Appl Neurobiol 2:439–450

Goy RW, McEwen BS 1980 Sexual differentiation of the brain. MIT Press, Cambridge, Massachusetts

Herschkowitz N 1988 Brain development in the fetus, neonate and infant. Biol Neonate 54:1–19

Legrand J 1980 Effect of thyroid hormone on brain development, with particular emphasis on glial cells and myelination. In: Di Benedetta C et al (eds) Multidisciplinary approach to brain development. Elsevier/North-Holland, Amsterdam p 279–292

Lewis PD, Balazs R, Patel AJ, Johnson AL 1975 The effect of undernutrition in early life on cell generation in the rat brain. Brain Res 83:235–247

Lynch A 1973 Vulnerable periods in the developing brain. PhD thesis, University of Manchester, UK

McConnell P, Berry M 1978a The effects of undernutrition on Purkinje cell dendritic growth in the rat. J Comp Neurol 177:159–172

McConnell P, Berry M 1978b The effect of refeeding after neonatal starvation on Purkinje cell dendritic growth in the rat. J Comp Neurol 178:759–772

Otake M, Schull WJ 1984 In utero exposure to A-bomb radiation and mental retardation; a reassessment. Br J Radiol 57:409–414

Patel AJ, Balazs R, Smith RM, Kingsbury AE, Hunt A 1980 Thyroid hormone and brain development. In: Di Benedetta C et al (eds) Multidisciplinary approach to brain development. Elsevier/North-Holland, Amsterdam, p 261–277

Peeling AN 1982 Quantitative histological investigation of effects of undernutrition on development of rat visual cortex. PhD thesis, Manchester Polytechnic (CNAA), UK

Raisman G, Field PM 1973 Sexual dimorphism in the neuropil of the preoptic area of the rat and its dependence on neonatal androgen. Brain Res 54:1–29

Rodier PM 1980 Chronology of neuron development: animal studies and their clinical implications. Dev Med Child Neurol 22:525–545

Rosenzweig MR, Bennett EL 1978 Experiential influences on brain anatomy and chemistry in rodents. In: Gottlieb G (ed) Studies on the development of behavior. Vol 4: Early influences. Academic Press, New York, p 289–327

Smart JL 1990 Vulnerability of developing brain to undernutrition. Uppsala J Med Sci Suppl 48: 21–41

Spalding DA 1873 Instinct with original observations on young animals. Macmillan's Magazine 27:282–293; reprinted 1954 Br J Anim Behav 2:2–11

Stewart CA 1918 Changes in the relative weights of the various parts, systems, and organs of young albino rats underfed for various periods. J Exp Zool 25:301–353

Stockard CR 1921 Developmental rate and structural expression: an experimental study of twins, 'double monsters', and single deformities and their interaction among embryonic organs during their origins and development. Am J Anat 28: 115–275

Thorpe WH 1963 Learning and instinct in animals, 2nd edn. Methuen, London

Warren MA, Bedi KS 1984 A quantitative assessment of the development of synapses and neurons in the visual cortex of control and undernourished rats. J Comp Neurol 227:104–108

Wiggins RC 1982 Myelin development and nutritional insufficiency. Brain Res Rev 4:151–175

Yakovlev PI, Lecours AR 1967 The myelogenetic cycles of regional maturation of the brain. In: Minkowski A (ed) Regional development of the brain in early life. Blackwell, Oxford, p 3–70

DISCUSSION

Dobbing: I would like to know measures of outcome after malnutrition other than anatomical ones, such as behavioural correlates, if there are any, and to what extent they occur. For me, this is most important.

As regards the differential susceptibility of the cerebellum to undernutrition, we should stress that if the cerebellum is growing at the same time as the rest of the brain but faster than it, and therefore over a shorter period, it is differentially sensitive in the sense that undernutrition at that particular time will affect the cerebellum more. It is not that the *process* of cerebellar growth is any more sensitive, but, as Jim Smart clearly said, it is affected more by virtue of its different growth characteristics. However, in practical terms, as he showed us, if you undernourish a developing rat at a particular time after birth when its cerebellum is growing fast and the rest of the brain is growing too, but more slowly, in those circumstances of a contemporaneous stress, there will be an overall differential deficit in the cerebellum. Again what interests me most about this is the degree of correlation of growth restriction with behavioural outcome.

Blakemore: It is an attractive notion that periods of sensitivity and vulnerability correspond only to periods of normal rapid development, and that, once the growth of a structure is complete, it's no longer either sensitive or vulnerable. In general that appears to be true, particularly for irreversible, unrepeatable processes, such as the generation of neurons and developmental cell death. But there are now important examples of more subtle modulations of brain function by chemical or hormonal influences, or by the external environment, at a stage outside any obvious spurt in growth, which don't correspond to any obvious morphological change. Such effects clearly depend on the presence of modifiable synaptic connections between cells, which can be exploited to produce subtle changes in local neuronal circuitry.

To take an extreme example, we are capable of forming memories at virtually any age; this ability is thought to depend on changes in the strength of synaptic connections in the hippocampus and/or the neocortex, where cells complete their differentiation and their obvious anatomical maturation relatively early in life. So, I don't think the sensitivity of the brain is restricted to periods of obvious growth.

Dobbing: I agree; but that has never been proposed. The vulnerable period hypothesis is not meant to account for all varieties of vulnerability.

Smart: There may be a distinction to be made here. I would regard the hypothesis that I was discussing as relating specifically to brain growth. I think what Colin Blakemore is talking about is a finer adjustment and modulation than that, which I wouldn't consider growth at all. So I don't see any conflict here.

Hanson: You stressed that lack of good postnatal nutrition is potentially more harmful for the offspring than prenatal nutrition, Dr Smart. If that is generally true, in teleological terms, why have more mammalian species not evolved the strategy of being precocial, like guinea pigs?

Smart: One could suggest that the events that occur before birth in brain development in most mammals are nevertheless the more important events: for example, the vast majority of neuron production in mouse brain occurs prenatally. It may be that what happens postnatally in brain growth in most altricial species, like rat, mouse, or man, is rather less important. There would not be the evolutionary pressure for those events to occur *in utero*.

Hanson: The other side of the coin is the question of whether the guinea pig is 'safe', once it is born, from nutritional disturbances. Since it can be weaned on the day of birth, to what extent can you alter its development by altering its diet experimentally?

Smart: Yes, probably it is relatively safe in terms of the gross aspects of brain growth and development. However, I must admit that I know of no study of postnatal undernutrition in guinea pigs.

Dobbing: I don't think anyone has explained satisfactorily, even in teleological terms, the species differences in maturity at birth. They have been said to be related to differences between nesting and non-nesting species, and the greater need of the latter to be kept warm and mobile, for which they must be more mature. This is not protection in the sense being discussed here.

Lucas: On the point of relating structure to function, could you tell us, Dr Smart, whether you would have predicted the data that I presented in my paper, in relation to your sequences of events? We have observed major deficits in motor development at 18 months in preterm infants as a result of a period of nutritional deprivation, if we like to call it that, during the third trimester principally. That could be explicable in terms of, say, damage to cerebellar development, but we also found significant reductions in cognitive development, which I wouldn't have predicted from your time sequences, in the sense that the major phase of neuronal development in most parts of the brain that might have been expected to relate to cognitive development would have been complete by the third trimester. Do you think that we picked up cognitive deficits simply because motor function is an important part of demonstrating cognitive skills in 18-month tests, or does this fit in with your ideas? And would you expect neurodevelopmental deficits at 18 months to be recoverable?

Smart: Your data may be consistent with the sequence of brain developmental events that I have outlined, in that not only does the cerebellum develop relatively late in the human (largely postnatally: Dobbing & Sands 1973), but so probably does the hippocampus. There are important changes happening in the hippocampus that might be consistent with cognitive effects. One tends to forget about the hippocampus because of the dramatic effects of undernutrition on the cerebellum, but in the dentate gyrus the granule cells are proliferating at the same time as the granule cells in the cerebellum, so you could perhaps be getting effects there.

Dobbing: I would have thought that a simpler explanation, that will be tested eventually, is that one of the defence strategies that you mentioned, namely the delaying of developmental processes, may be operating here. If you have a period of rapid development of many developmental processes and you delay them, it can be misleading to compare a retarded with a normally growing group cross-sectionally during a growth-spurt period. It can often happen that the differences you think you are showing are merely due to comparing populations cross-sectionally during a rapid phase of development, one of whose sigmoid growth curves is simply shifted to the right. It can be less misleading to make the comparison at a later, 'flatter' or more stable period of growth.

Parnas: Are there sensitive periods in cell migration in the human brain? In schizophrenia, this might be a part of the pathological process *in utero*, in the sense that noxious influences might disturb cell migration to their target positions and result in cytoarchitectural disarray.

Smart: Cell migration has not been closely studied in relation to undernutrition. I alluded to studies of the effects of thyroid hormones, and Balazs and his colleagues concluded that one of the effects of thyroid hormone, in either deficiency or excess, was to influence the rate of cell migration (Patel et al 1980). They postulated that the principal effects of thyroid hormone were mediated through effects on cell migration and/or differentiation. So cell migration can be affected in certain circumstances, by some factors. I don't know whether cell migration can be affected by undernutrition, but it wouldn't surprise me if it were.

Murray: You say that it is easy to forget about the hippocampus and concentrate on the cerebellum, but if one is interested in psychology, it's the other way round: the hippocampus is much more interesting than the cerebellum! Is it correct, if one can extrapolate from the rat to the human, that there may be a period postnatally in man when hippocampal development is still vulnerable to malnutrition?

Smart: Yes, almost certainly.

Murray: In that case, our paediatric colleagues ought to be able to measure the volume of the hippocampus, using MRI scanning, in children who, as newborns, went through a period when their nutrition was poor.

Dobbing: The cerebellum is a circumscribed definable region of the brain, whereas the hippocampus is much less so. It cannot be dissected out in the dead specimen, and weighed reliably. There are many brain regions about which paediatricians and others might have good quantitative questions if they could be better defined anatomically.

Murray: In adults, you can measure the hippocampal volume using magnetic resonance imaging (MRI).

Casaer: The comments made so far suggest that the cerebellum is mostly involved in the coordination of postural and motor control. The role of the cerebellum as an initiator of fast and fine motor programmes in time and in space is at present well accepted and is closely related to the motor output of cognitive functions and thus of intelligence.

Blakemore: One should not reject the possibility that the cerebellum has something to do with cognitive behaviour, in view of evidence of cognitive disorders associated with cerebellar disease. However, we should be careful about interpreting any gross and easily detectable anatomical changes associated with clinical conditions. Although MRI and other imaging techniques are very valuable, they may sometimes provide false leads. For instance, a few cases have been described in which severe autism is associated with cerebellar agenesis or atrophy, but this has not turned out to be a general rule and the relationship may not have been causal. One possible interpretation of these results is that growth of the cerebellum might be particularly vulnerable to (as yet unknown) deficits or disturbances that lead to autism, but that the functional defect that actually causes the symptoms of autism lies in some other brain structure (perhaps the cerebral cortex) whose growth is less obviously affected by the predisposing condition. The most exaggerated defects in brain structure, detected by imaging techniques, may not be responsible for the given functional disorder.

Wood: The commonest major lesion of the neonatal brain is intraventricular haemorrhage. This is a frequent event in low birth weight babies. Fairly minor degrees of bleeding with some extravasation into the cerebral substance appear to leave very little neurological deficit. That would not mean that there is no cognitive deficit, because very low birth weight babies tend to show long-term trends of poor school performance, measurable up to 7–8 years of age. Certainly there is a variability in the consequences of extravasation of blood around the ventricles.

Moxon: I just wanted to query the semantic distinction between growth and development, to be clear that the principles you were enunciating, Dr Smart, had to do with growth, and not development. If that is so, it implies an inherent quantitative aspect in growth which may not be present in development. You went on to discuss three possible compensatory mechanisms, one of which was changes in the kinetics of the cell cycle. How would you see that particular mechanism as a compensatory mechanism in the context of growth? I can see it in development, but I can't see it in growth.

Smart: Strictly, it not compensatory, because it is adjusting at the time of the growth—as growth is occurring—and therefore attenuating the effect of the prolonged DNA synthetic (S) phase of the cell cycle. The whole of the cell cycle is being changed at the same time, as far as one can judge; the S phase is being prolonged as a response to the undernutrition, but the G1 phase is being shortened, so it's all happening simultaneously. It would be best not called 'compensatory', which tends to imply something occurring after the events. It is happening at the time.

References

Dobbing J, Sands J 1973 The quantitative growth and development of the human brain. Arch Dis Child 48:757–767

Patel AJ, Balazs R, Smith RM, Kingsbury AE, Hunt A 1980 Thyroid hormone and brain development. In: Di Benedetta C et al (eds) Multidisciplinary approach to brain development. Elsevier/North-Holland, Amsterdam, p 261–277

Sensitive and vulnerable periods in the development of the visual system

Colin Blakemore

University Laboratory of Physiology, Parks Road, Oxford OX1 3PT, UK

Abstract. In advanced mammals the visual system consists of a number of parallel channels for the efficient processing of different aspects of the visual scene. Much of the basic anatomical structure of the visual pathway is constructed before birth. A wave of maturation sweeps through the system, from the eye to the visual cortex, the correct formation of connections depending on precisely timed interactions between axons and their targets. Competition between growing axons (apparently dependent on spontaneous impulse activity in those axons), cell death (partly influenced by competition between those cells' axons), axon withdrawal, trophic interactions—these and other mechanisms play a part in constructing the visual pathway and laying down basic 'maps' of the visual field before birth. Disturbances in such processes might underlie disorders of the genesis of the nervous system. At the level of the visual cortex, synaptic plasticity continues after birth and may permit cortical neurons to refine their processing capacities on the basis of information provided by the visual environment. This makes the young animal vulnerable to disturbances of visual experience early in life, which can cause virtually irreversible deficits in stereoscopic vision, visual resolution and sensitivity to contrast (amblyopia) in adult life.

1991 The childhood environment and adult disease. Wiley, Chichester (Ciba Foundation Symposium 156) p 129–154

The evolutionary origin of visual systems probably lies in the simple process of photosensitivity—the mere detection of the level of illumination in order to steer phototropic responses. Indeed, even in human beings, each photoreceptor in the retina does nothing more than catch light and signal its intensity. Yet from this simple process of point-sampling of the image flows the wealth of visual experience that dominates our knowledge of the world around us and underlies so much of our behaviour. Vision, the process of extracting meaning from the signals of photoreceptors, is an enormous task—perhaps the most complex piece of computation that the brain has to perform. We now know, from a combination of anatomical, neurophysiological and behavioural research, that the process of vision occupies a huge proportion of the cerebral cortex. In monkeys (and the situation is almost certainly very similar in humans), more than half the surface area of the cortex consists of a patchwork of perhaps

25 separate visual areas, each containing a representation of all or part of the visual field (Zeki 1975, 1990, Van Essen 1979, Kaas 1986), some of which seem to be dedicated to the analysis of particular aspects of the visual stimulus.

The active analysis of the visual image begins in the retina, which is, embryologically, an outgrowth of the forebrain and which contains a rich variety of neuronal types. The ganglion cells, whose axons make up the optic nerve, consist of a number of anatomically and functionally distinct classes, some responding to local brightening, others to darkening, some to static patterns, others better to movement, some to fine detail in the image, others to grosser features, and some (especially in primates) to the wavelength of light. Each separate class of ganglion cells is efficiently distributed across the retina: in general, the cell bodies of a particular class are arranged in a fairly regular mosaic and there is usually rather little overlap of the dendrites (and therefore the receptive fields) of neighbouring cells of the same class (see Wassle 1986).

The main impression to emerge from the past twenty years of anatomical and physiological study of the visual system, especially that of the monkey, is one of remarkable precision of organization. The optic nerve projects principally to the lateral geniculate nucleus (LGN) of the thalamus, where the fibres distribute themselves across the layers (typically three for each eye in Old World monkeys) to form remarkably exact 'maps' of the opposite half of the visual field. Indeed, the number of relay neurons in the LGN is similar to the number of fibres in the optic nerve and the major input to each LGN cell seems usually to be provided by a single optic nerve axon. Thus the receptive fields of geniculate cells are similar in arrangement to those of ganglion cells: they consist of a circular *centre*, which responds to brightening (for about half the population) or darkening (for the other half), with a concentric *surround* that is antagonistic to the centre. At any point in the visual field, the dimensions of the centres of the receptive fields of geniculate cells are similar to those of the retinal ganglion cells that provide their input.

The determination of visual acuity

The size of the centre of a cell's receptive field determines the finest detail in the visual image that it can detect. In the middle of the monkey or human fovea, where visual acuity is highest, there appears to be preservation of information from individual foveal cones, with many of the midget ganglion cells having receptive field centres consisting of just one cone and passing that single-cone signal to one LGN cell. In neurophysiological experiments on anaesthetized Old World monkeys, Blakemore & Vital-Durand (1986a) stimulated the receptive fields of LGN cells with high-contrast drifting gratings (patterns of alternating light and dark bars) and found the finest grating to which each cell would just respond (essentially measuring the visual 'acuity' of individual neurons). Their results (summarized in Fig. 1) are compatible with the hypothesis that some cells

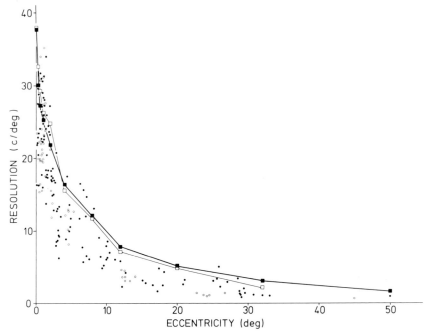

FIG. 1.   Results of Blakemore & Vital-Durand (1986a) showing the relationship between
human visual acuity in different parts of the visual field and the ability of individual
neurons in the lateral geniculate nucleus (LGN) of Old World monkeys to detect gratings.
The graph plots acuity in terms of the highest number of cycles of a high-contrast grating
(each cycle is one light plus one dark bar) that could just be resolved. Each small circle
(filled for the left eye, unfilled for the right) shows the 'acuity' of an individual cell
in the LGN of a relatively mature monkey (more than 6.5 months old), determined by
presenting a drifting grating to the receptive field and making it finer and finer until
the cell just stopped responding to the bars of the grating. The abscissa plots the position
of the cell's receptive field in the visual field, zero being the projection of the fovea.
The squares (filled for the left eye, unfilled for the right) plot determinations of acuity
of a human observer viewing drifting gratings on the same screen with different parts
of his retina. At each point in the visual field, the best LGN cells were as good at resolving
gratings as was the human observer.

in the central foveal representation are dominated by input from single
cones and that signals from these individual neurons may determine the
perceptual ability of the whole animal. The best-resolving cells could detect the
presence of gratings up to nearly 40 c/deg (cycles per degree of visual angle),
which is equivalent to a value of visual acuity superior to the clinically
accepted performance on Snellen acuity letters of a normal human observer
(6/6 or 20/20). This value is also close to the expected ultimate limit to
spatial resolution set by the angular dimensions of foveal cones (Jacobs &
Blakemore 1988).

At the level of the primary visual cortex, to which the LGN projects, there is also a retinotopic 'map' of the opposite visual half-field, though it is sloppier than that in the LGN and the sizes of receptive fields of cortical neurons are generally larger than those of the geniculate cells that provide their input. The majority of cortical cells respond best to an elongated bar or edge falling on their receptive fields, and they are selective for the orientation of the contour (Hubel & Wiesel 1977). Nevertheless, some neurons in the foveal representation of the visual cortex behave as if their ability to resolve fine detail in the image depends on input from single cones (or a row of single cones). They respond to drifting gratings (of the appropriate orientation if they are orientation selective) up to a resolution limit similar to that of the best LGN cells (see Blakemore 1990). They can also be remarkably efficient in their ability to detect contrast (the difference in intensity between neighbouring areas of the retinal image): when stimulated with a grating whose contrast is varied, some of them have a threshold indistinguishable from that of a human observer viewing a comparable visual stimulus (Hawken & Parker 1990).

Given the complexity and precision of organization of the adult visual system, it is not surprising to discover that it is vulnerable during its early development. We now know a good deal about both the prenatal and the postnatal maturation of the visual pathway in advanced mammals, and such work has revealed a variety of critical phases in development at which metabolic, hormonal and nutritional deficiencies might influence the system. Most remarkable is the overwhelming evidence that the visual pathway (and especially the visual cortex) is vulnerable to disturbances of visual experience itself for some time after birth, and that the defects in development that result from such disturbances can have a permanent effect on visual performance.

Prenatal development of the visual pathway

A wave of development sweeps through the visual pathway during gestation, from retina to cortex. In rhesus monkeys, for which the gestation period is about 165 days, Rakic (1977) has used cell birthdate labelling techniques to show that the first retinal ganglion cells are generated about the 30th day (E30), while those destined to form the LGN start to be born around E36. Neurogenesis of the cells of the primary visual cortex extends from around E43 to E102 and all the neurons that make up the pathway to the cortex are in place some two months before birth. Cortical neurons are generated from cells of the proliferative neuroepithelium near the surface of the lateral cerebral ventricle and they migrate up along the processes of radial glial cells towards the surface of the cerebral wall. The firstborn cells form a layer called the *preplate*, most of whose neurons are destined to die before or shortly after birth. Further waves of migrating juvenile neurons penetrate the preplate, dividing into an upper layer, which becomes the cell-sparse layer 1 of the mature cortex, and the *subplate*, a cell population that lies below the cortical plate proper, in what will become the white matter (see O'Leary 1989). The layers of the cortex itself are formed in an inside-out sequence, each successive generation of immature neurons migrating through the existing layers to take its place under layer 1 at the top of the plate. In rodents, despite the much shorter gestation period (22 days in the rat), all neurons appear to be generated by birth, but migration of cortical cells is not complete until the end of the first postnatal week.

Cell death, exuberance and competitive reorganization

The development of the visual system (and of many other parts of the brain) proceeds, in part, through initial overproduction of neurons and overconnection of axons, followed by phases of regression to establish the mature pattern (see Innocenti 1981, Cowan et al 1984). In the retina, for instance, many more ganglion cells are born and send out their fibres into the optic nerve than survive in the adult. In monkeys and cats more than half of all ganglion cells die before birth: some of this elimination of neurons may be genetically preprogrammed, but components of it are regulated on the basis of *competitive interaction* between the dendrites of neighbouring ganglion cells in the retina (Perry 1989) or between their axons in the central structures to which they project (Shatz & Sretavan 1986).

In both the LGN and the superior colliculus (the major midbrain visual centre, which receives a substantial input from some classes of ganglion cells) optic nerve axons from *both* eyes are initially intermingled throughout the entire structure when they first invade early in gestation. Gradually the left-eye and right-eye terminals segregate from each other to form separate layers (and, in the colliculus, to cover different fractions of the surface area of the nucleus: see Insausti et al 1985). This process of ocular segregation is particularly striking in the LGN, whose cells form a more-or-less homogeneous mass for some time after they have initially migrated into position and innervation by optic nerve fibres has commenced (up to about E83 in the rhesus monkey: Rakic 1977). The laminated structure of this nucleus, which is such a distinctive feature in the adult, emerges gradually over the final third of gestation. In the human fetus too, layers first start to appear in the LGN during the sixth month of gestation (Preobrazhenskaya 1965).

As the cells of the LGN redistribute themselves to form layers, so the left-eye and right-eye axons separate from each other to provide monocular input to each layer. The whole process is driven by 'competitive' interaction between the two sets of optic nerve fibres, since removal of one eye at a stage before the formation of layers (E64 in the rhesus monkey) leaves the nucleus quite uniformly innervated by the remaining eye, and its neurons do not become redistributed to form layers (Rakic 1979). Significantly, early removal of one eye in a fetal monkey also prevents a fraction of the expected death of ganglion cells in the remaining eye (Rakic & Riley 1983). All this implies that the final efficient distribution of ganglion cell types across the retina and the precise pattern of termination of their axons in central structures are achieved by initial overproduction followed by selective elimination.

Recent evidence suggests that spontaneous impulses propagating along the fibres of the optic nerve regulate the process of competition between their terminals. Sretavan et al (1987) have shown that, even with both eyes intact, the arborizations of optic nerve axons within the LGN fail to become restricted to their normal laminar pattern in fetal cats if the drug tetrodotoxin (which blocks action potentials) is applied continuously to the optic tract through the period over which segregation should occur.

Guidance of projections from thalamus to cortex

At early stages in its development, the cerebral cortex is an apparently uniform plate, but after all its neurons are in place it becomes differentiated into a large number of distinct areas, which can be classified according to their histological appearance, their patterns of afferent and efferent connections, and by their functional specializations. In particular, each sensory cortical area has its major afferent input from a corresponding specific relay nucleus in the thalamus (e.g. the LGN for the primary visual cortex). There is currently great interest in the possibility that these committed cortical areas emerge from an undifferentiated *protocortex*, which is more-or-less equipotential (see McConnell 1989, O'Leary 1989). If this were the case, it might simplify the huge problem of genetic specification of such a complex structure, because it would mean that the differentiation of each cortical zone is not pre-specified directly at the birth of the neurons in the germinal epithelium but is imposed on the equipotential cortex by some other inductive influence. One obvious candidate for such an inducing agent is the specific thalamic input that each area receives at around the time that its distinct histological appearance begins to appear.

Very recently, Molnar and I (Blakemore & Molnar 1991) have been using techniques of organotypic culture of neural explants in serum-free medium (Romijn et al 1988) to study the factors involved in determining the selective innervation of a given cortical area by its appropriate thalamic nucleus. We co-cultured together small explants of specific regions of the fetal rat thalamus (usually taken at E16, the time of maximum outgrowth of axons *in vivo*) and of postnatal cerebral cortex. In the normal rat, fibres from the LGN accumulate below the occipital cortex by the time of birth and invade the cortical plate during the first postnatal week (Lund & Mustari 1977). Yamamoto et al (1989) have already shown that an explant of E16 posterior thalamus, containing the LGN, will send axons into an explanted slice of neonatal visual cortex and form functional synaptic connections.

We have explored the effects of the age of the cortical explant, with results that imply that cells in the cortex may express a number of trophic influences on thalamic axons at different ages. With newborn cortex, between the day of birth and about the fourth postnatal day (P0–3), axons from a nearby E16 thalamic explant invade the cortical slice at about 1 mm per day, following a remarkably normal pattern of radial ingrowth. However, they do not stop, branch and terminate principally in the middle of the thickness of the cortex, as thalamic axons do in all cortical areas in the normal animal. They run up to layer 1 and some even ramify over long distances under the pial surface, with growth cones clearly visible at their tips. This pattern of ingrowth presumably depends on some general attractive influence (presumably a diffusible trophic substance) that the visual cortex has on the thalamus. This trophic influence is certainly not universal because thalamic axons are very reluctant to invade a similar explant of cerebellar cortex (which they do not innervate *in vivo*).

If the slice of visual cortex is somewhat older, taken between about P5 and P10, the invasion of thalamic axons is different: they run in radially, but most branch and appear to form terminals, after 3–4 days of culturing, in the middle of the cortical slice, presumably forming synapses in layer 4, as in the normal animal. Thus it seems that the cells of layer 4, many of which are still migrating through the cortical plate into position immediately after birth (Lund & Mustari 1977), might express a specific signal from about the fourth postnatal day, which causes ingrowing thalamic axons to terminate.

We had fully expected that these trophic influences of cortex on thalamus in culture would be selective for the *position* from which the two explants were taken and that they might therefore account for the guidance of fibres from particular thalamic nuclei to their correct cortical target areas in the normal animal. To our surprise, we have found that an explant of *any* part of the fetal thalamus will innervate any part of the cortex with identical patterns of ingrowth (depending only on the age of the cortical slice). Even if an explant of, say, LGN is given a choice by being co-cultured with a slice of occipital (putative visual) cortex on one side and a slice of frontal (putative sensorimotor) cortex on the other, it will send axons into both with indistinguishable patterns and densities of innervation. Thus, it appears that the entire immature cortex expresses only general trophic influences on thalamic axons, without any positional specificity (at least *in vitro*).

How then do thalamic axons find their way to the appropriate cortical targets with such uncanny precision in the intact animal? One possibility that emerges from recent observations by McConnell et al (1989) in the cat is that the correct routes for growth are established by a pioneering set of axons that grows downwards at a very early age from the subplate cells of the cortex to subcortical structures, including the thalamus. If these pioneering fibres were to be guided from each particular cortical site to the appropriate thalamic nucleus by some positionally selective chemical influence of the thalamic target on the subplate cells, they might establish axon pathways which could then be used for guidance by the ascending thalamo-cortical axons. Using techniques to label axons fluorescently in the whole, fixed fetal brain, Molnar and I have shown that, in the rat, as in the cat, there is indeed an early cortico-thalamic outgrowth, which starts to leave the occipital cortex on E14, and these axons are directed at corresponding targets in the thalamus. However, when we co-cultured fetal explants of cortex and explants of thalamus, axons did indeed grow out of the cortex (presumably from subplate cells) and invade the thalamus, but there was again no hint of any positional preference: any region of subplate will innervate any region of thalamus *in vitro*.

The mechanism of guidance of both cortico-thalamic and thalamo-cortical fibres remains obscure at the moment, but each of them may guide the other over the final stage of its growth. They seem to start their growth at roughly the same time, and to meet each other in the primitive internal capsule: then each set of fibres may fasciculate on the proximal portion of its counterpart to find its way to its target in the cortex or the thalamus. It would follow that the establishment of the *initial* outgrowth of each projection must depend on some inherent characteristics of the environment through which the axons grow. It is important, though, to emphasize that the brain is very small and the structural relationships between telencephalon and diencephalon are very simple at the stage at which these pathways are established. Axons may merely have to take the shortest available route, avoiding crossing their neighbours, in order to lay down specific tracks for their counterparts from the corresponding structure to grow along. Subsequent growth of the forebrain rapidly transforms the appearance of these early, simple fibre bundles into tortuous cables. This points up the critical importance of *timing* in the development of the nervous system. Many trophic interactions and the establishment of axonal pathways can occur only at specific times, which must surely make the system vulnerable to the possibility of disruption.

Afferent and efferent pathways of the visual cortex

By the time of birth in the cat, the major projections between the primary visual cortex and subcortical structures are present, if not complete, and they are at least roughly topographically organized (Henderson & Blakemore 1986). All this must have been achieved without the benefit of any direct visual stimulation because the outer segments of the photoreceptors do not become functional until some days after birth in the cat. This is not to say, however, that impulse activity in nerve fibres is not involved. There could be spontaneous activity originating in the retina and possibly correlated for neighbouring ganglion cells, which could play a part in coordinating innervation and in 'map' formation, just as it seems to regulate competitive interaction in the developing LGN.

Postnatal development of the visual pathway

Despite the superficial appearance of anatomical completeness of the visual system at birth in higher mammals, and the fact that neurogenesis and long-range axon-outgrowth have stopped, a great deal of important refinement of local connectivity takes place after birth. At least some of these postnatal maturational changes are dependent on visual stimulation.

Exuberance, axon withdrawal and cell death in the development of cortico-cortical connections

Many of the connections between cortical areas, on the same side of the brain and between the two hemispheres, are remarkably florid and exuberant in their organization early in life in cats and rodents (though they seem to be more mature in the monkey). For instance, in the adult, the body of the primary visual cortex contributes few if any axons to the corpus callosum and the interhemispheric projection mainly comes from (and projects to) a strip of cortical territory immediately adjacent to the primary visual cortex: but in the newborn animal the primary visual cortex on the two sides is extensively interconnected. The elimination of this exuberant pathway is accomplished by the withdrawal of axons from the corpus callosum without the death of the cells of origin (Innocenti 1981), and the extent of the postnatal reshaping of this projection can be influenced by unusual visual experience, such as strabismus (Lund et al 1978). Recently, LaMantia & Rakic (1990) have shown that, despite the fact that the pattern of interhemispheric connections is qualitatively mature at birth in the monkey, 70% of the axons in the corpus callosum disappear over the first few postnatal months, indeed at a rate of 50 axons per second during the first three weeks. It is conceivable that the condition of agenesis of the corpus callosum in humans is due to a failure to control the elimination of callosal axons, not just a failure to lay them down.

Many of the cortico-cortical projections within the same hemisphere are also aberrant in origin and exuberant in termination for some time after birth, at least in the cat. For example, the primary visual cortex even receives a transient projection from the auditory area of the cortex early in life (Dehay et al 1985). Price & Blakemore (1985) found that the projection from the primary to the neighbouring second visual area is initially aberrant in that the cells of origin are equally dense in the lower and the upper layers of the primary cortex (whereas in the adult they lie mainly in the superficial layers), and that there is no evidence of clustering of these cells in a mosaic of clumps, as in the adult. The adult pattern emerges in the week after eye-opening but this basic process does not seem to require visual experience. Interestingly, the loss of the projection from the deep-layer cells seems to be achieved largely through their death, whereas the creation of the upper-layer clusters seems to depend on the withdrawal from the second visual area of the axons of cells lying between the clusters, without their actual death.

The basis of amblyopia

The most direct evidence that the vision of human beings can be disturbed by abnormal visual experience early in life comes from the condition of *amblyopia*, a deficiency of vision, usually in only one eye, that follows either unilateral *deprivation* of vision (through cataract, ptosis of the lid, etc.), *anisometropia* (relative defocus of the image in one eye) or *strabismus* (particularly inward deviation of one eye) early in life. Amblyopia is defined clinically as a reduction of visual acuity (sometimes to a level below the legal definition of blindness) but it often also involves poor sensitivity to contrast, loss of stereoscopic vision and other more complicated disturbances of visual perception (see Hess et al 1990).

Until recently, although this condition is very common (affecting perhaps 3% of the population) and had been extensively described clinically, the basis of the deficiencies was unknown. Over the past 25 years or so, anatomical and neurophysiological studies in cats and monkeys (which also suffer amblyopia after the same kinds of early visual disturbance) have produced a growing understanding of the basis of the defects of acuity and contrast sensitivity that follow deprivation or anisometropia.

Although prolonged deprivation or artificial paralytic strabismus (produced by removal of the lateral rectus muscle) can produce changes in the anatomy and physiology of retinal ganglion cells, evidence points to the primary visual cortex as the main site of the defect in visual acuity. Hubel & Wiesel showed that temporary closure of the lids of one eye in kittens causes neurons in the visual cortex, the majority of which can be excited through either eye in the normal adult, largely to become unresponsive through the deprived eye (see Hubel & Wiesel 1970). Figure 2 plots these results to show the way in which the sensitivity of cortical neurons to the effects of monocular deprivation rises rapidly after eye-opening, reaches its peak at four weeks of age (when as little as a day or two of deprivation can virtually eliminate input to the cortex from the deprived eye) and declines gradually over the following two months.

This functional disconnection is accompanied by actual anatomical changes in the distribution of afferent fibres from the right-eye and left-eye layers of the LGN in layer 4 of the cortex. In both cats and monkeys a process of ocular segregation, very similar to the competitive interaction that generates the layers of the LGN during gestation, takes place *after* birth in the visual cortex. Although geniculate fibres start to arrive in the fourth layer before birth, the terminals of left-eye and right-eye axons are initially intermingled. The pattern of terminal distribution in the cortex, which can be revealed by the technique of transneuronal autoradiography after injection of labelled proline into the vitreous body of one eye (see Swindale et al 1981), gradually undergoes ocular segregation during the first two months of postnatal life. In the normal adult monkey, LGN axons carrying signals from one eye terminate in a pattern of

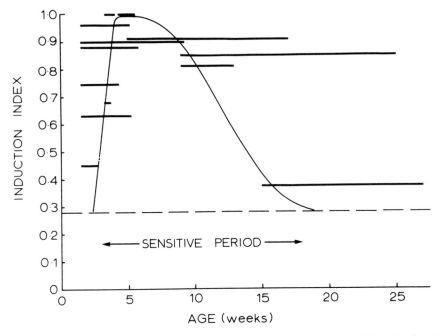

FIG. 2. Physiological effects of monocular deprivation on neurons of the visual cortex of cats (see Hubel & Wiesel 1970), plotted graphically to reveal the period of sensitivity. Recordings were always made in the hemisphere opposite the deprived eye and the relative excitability of cortical neurons through the two eyes was determined. The induction index is the ratio of the number of cells dominated by the ipsilateral (non-deprived) eye to the total number of cells studied. The interrupted line is the index calculated for a large sample of cortical cells from normal cells: about two-thirds of cells are dominated by the opposite eye in the cat cortex. Each horizontal bar indicates, for one monocularly deprived animal, the duration for which the eye was closed. (Note that the eyes do not normally open until about 10 days of age.) Very short periods of monocular deprivation around four weeks of age cause virtually all cells to become dominated by the open eye, but even prolonged deprivation after three months has little effect on ocular dominance.

surface-parallel bands, each about 300 μm across, with gaps in between, of the same width, which are occupied by similar bands of terminals from the other eye. This alternating pattern emerges by a process of competitive interaction, which is dramatically influenced by any imbalance of activity in the two sets of fibres. Thus, after monocular deprivation, the terminals belonging to the deprived eye become restricted to shrunken patches, while those from the other eye fill the enlarged space in between (Fig. 3A). Presumably this failure in anatomical innervation at least partly explains the loss of functional input to cortical cells from a deprived or defocused eye.

   The conventional treatment for amblyopia involves patching of the good eye, but this procedure is usually effective only in young children and especially in

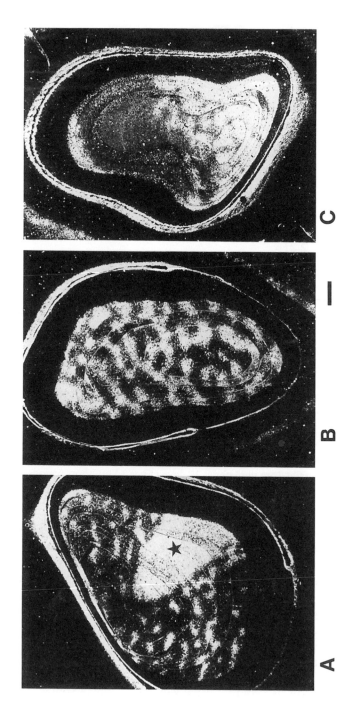

140

FIG. 3. Swindale et al (1981) used the transneuronal autoradiographic technique to reveal the pattern of termination in the monkey's primary visual cortex of lateral geniculate nucleus (LGN) axons carrying signals from one eye. The vitreous body of the right eye was injected with tritiated proline and 11–19 days later the animal was perfused with 4% paraformaldehyde and surface-parallel sections of the visual cortex were cut on a freezing microtome. Sections were coated with photographic emulsion, exposed for 2–3.5 months and developed to reveal areas of high radioactivity. These are montages of photomicrographs of sections through the calcarine region of the cortex, which represents the more peripheral visual field: the central portion of each montage shows the lower part of layer 4, where the bulk of the LGN axons terminate. The sections were photographed with dark-field illumination, which shows areas of high radioactivity, concentrated in the terminals of axons from the LGN layers connected to the injected right eye, as bright. Because of the curved shape of the calcarine cortex, radioactively labelled white matter appears as a ring at the periphery of each montage.

A. Right hemisphere of an animal monocularly deprived in the right eye from two to 23 days and studied at that age. Even such a short period of deprivation had caused the terminals of the deprived eye to occupy only small patches, except in the large oval region marked with a star, which presumably corresponds to the optic disc of the contralateral, left eye, and which is filled by terminals from the right eye because it has no input from the left eye and therefore has no competitive interaction.

B. Right hemisphere of a monkey initially deprived in the right eye until 26 days, at which stage the right eye was reopened and the left eye closed for a further six days before study. Now, after just six days of competitive advantage, the right eye's terminal distributions have expanded to form relatively normal bands occupying about half the area of layer 4.

C. Left hemisphere of an animal deprived until 28 days and then reverse-deprived for 15 days. The dark area, free of terminals from the right eye, presumably corresponds to the optic disc of that eye and is entirely filled by left-eye input. Elsewhere, the input from the originally deprived right eye has expanded virtually to fill the whole extent of layer 4.
Scale bar, 1 mm.

babies less than, say, two years old, where patching can be used in conjunction with non-verbal methods of assessment of visual acuity (see Boothe et al 1985) to monitor its effects. In the same way, reverse deprivation of cats and monkeys (opening a deprived eye and closing the other) early in life, when the animal is still within the sensitive period, can reinstate functional input to cortical cells from the initially deprived eye and can cause the bands of terminals from that eye to expand (see Swindale et al 1981 and Fig. 3).

Activity-dependent redistribution of LGN terminals in the primary visual cortex is accompanied by parallel changes in the appearance of their cells of origin in the LGN. Monocular deprivation leads to a failure of normal cell growth in the LGN layers receiving from that eye (Fig. 4A) and early reverse deprivation can restore the imbalance of growth. One interpretation of these results is that the growth of the soma of a neuron depends on the success of its axons in arborizing and establishing terminals (see Swindale et al 1981). In those terms the dramatic effects of deprivation on the LGN would be *secondary* to the results of competitive interaction in the cortex.

The correlation between amblyopia in humans and changes in the anatomical and physiological dominance of the visual cortex caused by monocular deprivation is impressive, but the latter results do not directly account for the reduction of acuity in amblyopia. A more complete account of the human disorder has been obtained from experiments in which the actual 'acuity' of individual neurons has been determined by finding the finest grating pattern to which they will respond (see Fig. 1).

Such experiments have shown that the best-resolving cells of the foveal representation in the LGN are very poor in 'acuity' at birth and undergo a seven-fold improvement during the first year or so of life in normal monkeys (Blakemore & Vital-Durand 1986a). At every stage, the best cells in the cortex are similar in performance to the best in the LGN. Behavioural experiments have shown that visual acuity is indeed very low at birth in both monkeys and human babies and that it normally increases slowly with age (see Boothe et al 1985). Interestingly, it appears that many of the maturational changes in neuronal properties are accounted for not by reorganization of connections somewhere between photoreceptors and LGN cells but simply by growth of the eyeball (which increases the magnification of the image), combined with the fact that the fovea actually *shrinks* in diameter and the individual foveal cones become considerably more slender and more closely packed (Jacobs & Blakemore 1988, Blakemore 1990).

Deprivation of vision has very different effects on the development of 'acuity' of LGN and cortical neurons. Despite the gross anatomical disturbance in the LGN produced by monocular deprivation, neurons are easily recorded in the deprived layers (Fig. 4B) and their physiological properties are remarkably normal (Blakemore & Vital-Durand 1986b). In particular, despite the fact that their receptive fields have never been stimulated with well-focused retinal images,

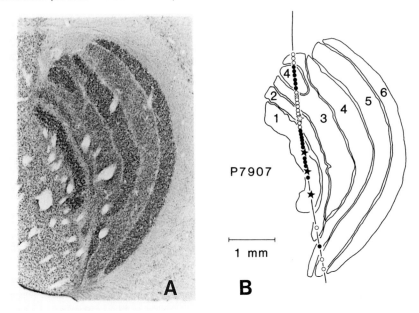

FIG. 4. A. Coronal section through the right lateral geniculate nucleus (LGN) of a monkey whose right eye had been closed from the day of birth until a physiological recording experiment at 189 days. Nissl staining reveals the three layers with input from the non-deprived left eye as densely stained. The three deprived-eye layers appear pale because the neurons are smaller and contain less Nissl substance.

B. A reconstruction of a microelectrode penetration through exactly this plane of the LGN of this animal. Along the course taken by the electrode, the circles and stars plot the positions at which LGN cells of different classes were recorded and studied. Filled symbols represent cells with receptive fields in the left eye and open symbols are deprived, right-eye cells. There was no difficulty in isolating cells in the deprived layers and their physiological properties were apparently quite normal. (Results of Blakemore & Vital-Durand 1986b.)

the best neurons recorded in the deprived layers have 'acuities' indistinguishable from those of the best cells in the normal layers. It seems that the profound amblyopia that can result from as little as two weeks of monocular deprivation in the monkey (see Boothe et al 1985) is unlikely to be due primarily to changes in the retina or the LGN.

On the other hand, deprivation has dramatic effects on the development of neuronal 'acuity' at the level of the visual cortex (see Blakemore 1990). After as little as a week of monocular deprivation around the fourth week, with subsequent binocular vision for more than a year, very few cortical cells respond through the deprived eye and those that do have poor ability to resolve gratings and detect their contrast (C. Blakemore & F. Vital-Durand, unpublished

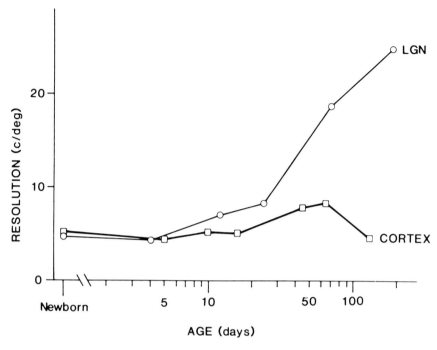

FIG. 5. This graph summarizes unpublished physiological results of Blakemore and Vital-Durand. Recordings were taken from the lateral geniculate nucleus (LGN) and the visual cortex of monkeys of different ages (plotted on the logarithmic abscissa). The resolution or 'acuity' of each cell was determined with drifting, high-contrast gratings (see Fig. 1) and this is plotted on the ordinate as the maximum number of cycles of grating per degree of visual angle to which the cell would just respond. All the data in this graph come from cells responding through eyes that had been continuously closed from the day of birth until the recording experiment. The results for the LGN (unfilled circles) came from the deprived layers of monocularly deprived monkeys, while those for the cortex (unfilled squares) came from binocularly deprived animals. Each animal is represented by the very best-resolving cell recorded in the foveal area (though it is important to emphasize that many other cells in each sample had similar values of acuity and therefore these best cells were representative of the sample).

observations). Deprivation of vision in *both* eyes eliminates competitive interactions, though it leaves the majority of cortical cells responsive through one eye or the other, not both. This allows one to study the effects of deprivation *per se* and the same result is seen: cortical cells in such animals have very poor 'acuity'. Fig. 5 is a comparison between the best-resolving cells found at different ages in the deprived layers of the LGN and in the cortex of binocularly deprived monkeys: it shows clearly the way in which neuronal performance improves in the LGN, apparently unhindered by a lack of normal visual stimulation, while it fails at the level of the cortex. This strongly implies that the pathological change underlying the reduction of acuity and loss of contrast sensitivity in deprivation (and anisometropic) amblyopia is a failure of first-order cortical cells to establish normal input from a small ensemble of geniculate axons.

Conclusions

The maturation of the visual pathway is a prolonged process, lasting until several years after birth in humans. A variety of mechanisms are involved in establishing the final pattern of functional connectivity: cell proliferation and migration; axon outgrowth guided by path-finding mechanisms of various sorts and by trophic influences from the target; overconnection followed by competitive interaction; regulation of connectivity by cell death and axon withdrawal. Many of these processes appear to be critically timed during prenatal or postnatal development and it is surprising that they so rarely seem to go wrong. The visual cortex undergoes a distinct sensitive period of 'plasticity' (or, more likely, a number of periods, different in duration for different properties) during the first few weeks, months or years of life in advanced, binocular mammals. It is clear that the synaptic input to cortical cells is being regulated on the basis of activity reaching the cortex from the eyes. This sensitivity to visual stimulation is presumably of adaptive importance to the animal, perhaps being needed to refine the mechanisms for the detection of fine detail in the image, for stereoscopic vision and for the identification of shape. However, this period of sensitivity also makes the cortex vulnerable to disturbances of visual experience, which can have permanent consequences for vision.

Acknowledgements

The research from my own laboratory described here has been supported by Programme grants from the Medical Research Council and by grants from the Wellcome Trust and the Science and Engineering Research Council. The work on the development of the primate visual system has largely depended on collaboration with F. Vital-Durand of INSERM Unité 94, Bron, France, supported in part by grants from INSERM and a Network Grant from IBRO.

References

Blakemore C 1990 Maturation of mechanisms for efficient spatial vision. In: Blakemore C (ed) Vision: coding and efficiency. Cambridge University Press, Cambridge, p 254–266

Blakemore C, Molnar Z 1991 Factors involved in the establishment of specific interconnections between thalamus and cerebral cortex. Cold Spring Harbor Symp Quant Biol 55, in press

Blakemore C, Vital-Durand F 1986a Organization and post-natal development of the monkey's lateral geniculate nucleus. J Physiol (Lond) 380:453–491

Blakemore C, Vital-Durand F 1986b Effects of visual deprivation on the development of the monkey's lateral geniculate nucleus. J Physiol (Lond) 380:493–511

Boothe RG, Dobson V, Teller DY 1985 Postnatal development of vision in human and nonhuman primates. Annu Rev Neurosci 8:495–545

Cowan WM, Fawcett JW, O'Leary DDM, Stanfield BB 1984 Regressive events in neurogenesis. Science (Wash DC) 225:1258–1265

Dehay C, Bullier J, Kennedy H 1985 Transient projections from the fronto-parietal and temporal cortex to areas 17, 18 and 19 in the kitten. Exp Brain Res 57:208–212

Hawken MJ, Parker AJ 1990 Detection and discrimination mechanisms in the striate cortex of the Old-World monkey. In: Blakemore C (ed) Vision: coding and efficiency. Cambridge University Press, Cambridge, p 103–116

Henderson Z, Blakemore C 1986 Organization of the visual pathways in the newborn kitten. Neurosci Res 3:628–659

Hess RF, Field DJ, Watt RJ 1990 The puzzle of amblyopia. In Blakemore C (ed) Vision: coding and efficiency. Cambridge University Press, Cambridge, p 267–280

Hubel DH, Wiesel TN 1970 The period of susceptibility to the physiological effects of unilateral eye closure in kittens. J Physiol (Lond) 206:419–436

Hubel DH, Wiesel TN 1977 Functional architecture of macaque monkey visual cortex. Proc R Soc Lond B Biol Sci 198:1–59

Innocenti GM 1981 Growth and reshaping of axons in the establishment of visual callosal connections. Science (Wash DC) 212:824–827

Insausti R, Blakemore C, Cowan WM 1985 Postnatal development of the ipsilateral retinocollicular projection and the effects of unilateral enucleation in the golden hamster. J Comp Neurol 234:393–409

Jacobs DS, Blakemore C 1988 Factors limiting the postnatal development of visual acuity in the monkey. Vision Res 28:947–958

Kaas JH 1986 The structural basis for information processing in the primate visual system. In: Pettigrew JD, Sanderson KJ, Levick WR (eds) Visual neuroscience. Cambridge University Press, Cambridge, p 315–340

La Mantia A-S, Rakic P 1990 Axon overproduction and elimination in the corpus callosum of the developing rhesus monkey. J Neurosci 10:2156–2175

Lund RD, Mustari MJ 1977 Development of the geniculocortical pathway in rats. J Comp Neurol 173:289–306

Lund RD, Mitchell DE, Henry GH 1978 Squint-induced modification of callosal connections in cats. Brain Res 144:169–172

McConnell SK 1989 The determination of neuronal fate in the cerebral cortex. Trends Neurosci 12:342–349

McConnell SK, Ghosh A, Shatz CJ 1989 Subplate neurons pioneer the first axon pathway from the cerebral cortex. Science (Wash DC) 245:978–982

O'Leary DDM 1989 Do cortical areas emerge from a protocortex? Trends Neurosci 12:400–406

Perry VH 1989 Dendritic interactions between cell populations in the developing retina. In: Finlay BL, Sengelaub DR (eds) Development of the vertebrate retina. Plenum Publishing Corporation, New York, p 149–172

Preobrazhenskaya NS 1965 Occipital area, lateral geniculate body, pulvinar and other subcortical structures of the visual analyser. (In Russian) In Sarkisov SA (ed) Development of child's brain. Medicina, Leningrad, p 87–127

Price DJ, Blakemore C 1985 Regressive events in the postnatal development of association projections in the visual cortex. Nature (Lond) 316:721–724

Rakic P 1977 Prenatal development of the visual system in rhesus monkey. Philos Trans R Soc Lond B Biol Sci 278:245–260

Rakic P 1979 Genesis of visual connections in the rhesus monkey. In: Freeman RD (ed) Developmental neurobiology of vision. Plenum Publishing Corporation, New York, p 249–260

Rakic P, Riley KP 1983 Regulation of axon number in primate optic nerve by prenatal binocular competition. Nature (Lond) 305:135–137

Romijn HJ, de Jong BM, Ruijter JM 1988 A procedure for culturing rat neocortex explants in a serum-free nutrient medium. J Neurosci Methods 23:75–83

Shatz CJ, Sretavan DW 1986 Interactions between retinal ganglion cells during the development of the mammalian visual system. Annu Rev Neurosci 9:171–207

Sretavan DW, Shatz CJ, Stryker MP 1987 Prenatal development of retinogeniculate axon arbors in the presence of tetrodotoxin. Soc Neurosci Abstr 13:591

Swindale NV, Vital-Durand F, Blakemore C 1981 Recovery from monocular deprivation in the monkey. III. Reversal of anatomical effects in the visual cortex. Proc R Soc Lond B Biol Sci 213:435–450

Van Essen DC 1979 Visual areas of the mammalian cerebral cortex. Annu Rev Neurosci 2:227–263

Wassle H 1986 Sampling of visual space by retinal ganglion cells. In: Pettigrew JD, Sanderson KJ, Levick WR (eds) Visual neuroscience. Cambridge University Press, Cambridge, p 19–32

Yamamoto N, Kurotani T, Toyama K 1989 Neural connections between the lateral geniculate nucleus and visual cortex in vitro. Science (Wash DC) 245:192–194

Zeki SM 1975 The functional organization of projections from striate to prestriate visual cortex in the rhesus monkey. Cold Spring Harbor Symp Quant Biol 40:591–600

Zeki S 1990 The motion pathways of the visual cortex. In Blakemore C (ed) Vision: coding and efficiency. Cambridge University Press, Cambridge, p 321–345

DISCUSSION

Thornburg: Can you tell us what determines the rate of firing of these neurons *in utero*?

Blakemore: I wish I could. If impulses in axons are regulating their growth and the way in which they form synapses *in utero*, it is important to know the nature of that activity, where it is set up, and how it is coordinated in a way that might guide the formation of connections. We know that in adult animals, the spontaneous activity of optic nerve fibres, recorded in the dark, tends to be correlated for ganglion cells lying close together in the retina. So, some inherent property of each part of the retina tends to make its neurons fire together spontaneously. This might be something as simple as the fact that the signal resulting from the photoisomerization of photopigment molecules in the receptor cells is amplified, so each photopigment molecule, isomerizing spontaneously, might set up a chain of amplified events, which increases the probability of synchronized firing in several neighbouring ganglion cells.

Thornburg: Do you assume that there is no light stimulation in the uterus?

Blakemore: In monkeys and humans there is surely little or no possibility of light stimulation in the uterus; but, in any case, whatever light is present must be diffuse and there could be no patterned retinal image. The postnatal influence of visual stimulation depends on pattern and contrast in the retinal image, not just the presence of light. Closing the eyelids of a monkey reduces mean retinal illumination by only half a log unit or so, because the lids are fairly translucent, but virtually eliminates contrast in the image. If one eye is closed in this way early in life there are changes in afferent innervation of the visual cortex and the excitability of cortical cells through that eye within a few days.

Lucas: One possible way for nutritional vulnerability to operate occurs to me. If different parts of the brain are developing at different times, could you imagine a situation in which nutritional deprivation allowed the whole brain to change its architecture, if a specific actively growing population of cells was not able to keep up for a while, and then, because the architecture of the brain had changed, those cells couldn't find their way to their appropriate sites? Is that a mechanism you could envisage?

Blakemore: That's a strong possibility. If a process is delayed, for nutritional or other reasons, then any change in the shape of the rest of the brain could prevent that process from happening at all, or interfere with it considerably. Even if neurons do not actually die or completely lose their capacity to grow, if they are prevented from sending out their axons at the right moment, things may go very wrong.

Lucas: Is there evidence in other situations for actual lesions in the brain resulting in neurons or axons ending up in the wrong place?

Blakemore: Damage to one part of the growing brain can certainly lead to the misrouting of axons that should normally innervate the lesioned area. This phenomenon has been used by D. Frost, M. Sur and their colleagues (see Sur et al 1990) to produce innervation of regions of cerebral cortex by inappropriate sensory axons. For instance, optic nerve axons in the developing hamster or ferret can be induced to innervate thalamic nuclei of the somatic sensory or auditory pathway by destroying the normal targets of the visual fibres (the lateral geniculate nucleus and the superior colliculus) and interrupting the ascending somatic sensory or auditory afferent pathway, respectively, below the level of the thalamus. When optic nerve fibres subsequently enter the wrong thalamic nucleus, they make synaptic contact with the cells there, whose axons in turn project up to the region of cortex appropriate for the interrupted sensory pathway. Thus visual input can be forced into the somatic sensory or auditory part of the cortex. Physiological recording from these aberrantly innervated regions of cortex has shown that some of the cortical neurons respond to visual stimulation and that they can have receptive fields similar to those normally found in the visual cortex itself. These experiments, like those involving co-culture of thalamus and cortex, which I have described, imply that the cerebral cortex is equipotential, to some extent, at early stages in development.

Wood: Professor Blakemore, it seems that Nature has been very careless in allowing a common disorder, namely squint, to interfere in such a sophisticated part of the development of the function of the brain as the development of vision.

Blakemore: True; but squint of this kind is only common in human beings and in primates that are protected from selective pressure by being reared in captivity. Some people have questioned the validity of animal models for studying the effects of strabismus, because squint is not seen in animals in the wild (except those with the albino mutation, such as the siamese cat, which squint for entirely different reasons). However, in captive populations of monkeys, protected in the same way as human populations, squint occurs naturally at about 4%—the same rate as in humans (Kiorpes et al 1985). So the tendency for squint is there but is selected against very strongly, presumably because of the visual disability it produces.

On the other hand, the kinds of reorganization of vision that result from squint early in life might be considered adaptive. A squinting animal has its eyes pointing in different directions; therefore, if normal binocular correspondence were to be maintained, the two eyes would provide conflicting information about the direction in space of any object of interest. This should result in double vision, which is indeed a common consequence of squint occurring later in life. The dramatic reduction in the proportion of binocular neurons in the visual cortex (producing independent monocular systems) and the common occurrence of amblyopia and suppression (essentially ignoring one eye) that result from early squint might resolve the problem of conflicting binocular information. However, opting for monocular vision carries the penalty of the loss of stereopsis—surely a severe disadvantage for arboreal monkeys, for which depth perception must be important for normal motor coordination.

Hamosh: We know that long-chain polyunsaturated fatty acids in the brain and retina are important for visual function, and for other brain functions. These fatty acids are deposited in the human brain in the last three months of gestation (Clandinin et al 1982); thus the brain of premature infants has lower levels of long-chain polyunsaturated fatty acids (LCPUFA) than that of full-term infants. LCPUFA are synthesized by elongation and desaturation of linolenic acid ($C18:3\omega-3$). These enzyme systems are however not developed in premature infants and their activity is low even in full-term infants (Clandinin et al 1982). Thus these nutrients have to be supplied by the diet. Elegant studies of Neuringer & Connor (1986) show that if rhesus monkeys are deprived of these fatty acids *in utero* and postnatally, their visual function is defective. One can replenish these fatty acids by providing them in the diet. However, while their concentration in the brain returns to normal, the functional deficit is not corrected (Neuringer & Connor 1986).

Blakemore: Are those long-chain fatty acids involved in the synthesis of myelin?

Hamosh: Some of the fatty acids are. The brain content of docosahexaenoic and eicosapentaenoic acids ($C22:6\omega-3$ and $C20:5\omega-3$ respectively) is high. We and others have shown that the milk secreted by mothers of premature infants has higher levels of these fatty acids than milk produced by mothers of term infants (Bitman et al 1983). Therefore if a baby is given human milk, it is advisable to feed gestationally comparable milk rather than full-term milk.

In the USA, the companies that produce the premature infant formulas are trying to supply these fatty acids in the diet, and several studies are currently going on, to evaluate the short- and longer-term effects of supplementation with LCPUFA.

Blakemore: One should not underemphasize the effect that interfering with myelination might have on normal functional development. The process of myelination could be significant in three ways during development. First, the myelin sheath simply separates axons from each other, which means that they can no longer participate in developmental processes that require close contact for the exchange of chemical or electrical signals between neighbouring axons.

Second, the classical function of myelin is, of course, to speed up impulse propagation, which would tend to increase the degree of synchronization of impulses arriving along different axons in target structures. The resulting improvement of correlation in firing might influence the selective strengthening of synapses.

The third point is that myelination seems to terminate the capacity of axons in the mammalian central nervous system to grow and to regenerate. Martin Schwab and his colleagues in Zurich have shown that oligodendrocytes, which synthesize myelin, manufacture at least two factors that specifically inhibit axon growth and prevent neurons in the mammalian brain from regenerating (Caroni & Schwab 1988). We don't know why the mammalian brain terminates its own capacity to regenerate, but it does so in association with the process of myelination.

Casaer: You discussed two types of 'pathfinders'; why are there two systems?

Blakemore: Because they are guiding different things. Immature *neurons*, before they produce axons, migrate along radial glial fibres from the germinal zone, where they are born, to their final positions in the cerebral or cerebellar cortex. In invertebrates, glial fibres can also form pathways along which axons grow towards their targets; but this is not thought to be a major method of guidance in the vertebrate brain. The growth and guidance of *axons* appears, rather, to depend on a variety of mechanisms. First, the general extracellular matrix contains a number of substances, including laminin, that simply facilitate axonal growth and hence make the environment permissive to growth. Fibres sometimes fasciculate on pre-existing pioneer axons (as I have suggested here for part of the growth of the thalamo-cortical projection) and specific adhesion molecules expressed on the surfaces of the fasciculating axons may play an important part in that process. And diffusible trophic substances, released by target structures, may attract growing axons, at a distance, towards their correct targets.

Casaer: It is tempting to draw a parallel to the human. In human development the first clear-cut new postnatal behaviour is the 'visual social smile' at the age of about eight weeks. In a longitudinal study, orientation-specific visually evoked cortical responses were shown to appear at exactly the same period, 8–10 weeks of postnatal age. Orientation-specific responses are thought, in our present state of knowledge, to be cortical and not subcortical (Atkinson 1989).

Blakemore: There is a paradox here that I find worrying. Although the gestation period is longer for humans than for monkeys, clinical evidence of amblyopia following monocular deprivation suggests that geniculo-cortical axons may not *begin* to segregate until about three months after birth in humans, whereas in monkeys the process begins before birth and is complete by about seven weeks postnatally.

Suomi: It is perhaps not so remarkable if you consider that the maturation rate of other developmental processes in rhesus monkeys is approximately four times that of humans.

Dobbing: The monkey is very much advanced at birth in terms of its brain growth compared to the human (Dobbing & Sands 1979). Interspecies differences in the timing of maturation in relation to birth need to be taken into account, as well as comparative rates of maturation (Dobbing 1990). They so rarely are!

Blakemore: We need direct anatomical information about the pattern of innervation from thalamus to cortex in humans, about when geniculate axons arrive at the visual cortex and when the competitive process between right-eye and left-eye fibres takes place.

Dobbing: The velocity curves of general brain growth in weight are almost certainly a guide. It's on that basis that one considers the monkeys to be much more developed at birth than man, as well as the quite clearly more mature behaviour of the newborn monkey.

Blakemore: The results of Atkinson & Braddick (1990), to which Professor Casaer refers, suggest that sensitivity to the orientation of lines and edges appears in the human visual cortex at about eight weeks after birth. At a similar age, stereoscopic mechanisms are beginning to develop, judged by behavioural criteria and from cortical evoked potentials to dynamic random-dot stereograms. At about the same time, sensitivity to monocular deprivation in human babies may just be appearing. All those things happen immediately after birth in monkeys. In the first week of life, the orientation selectivity of cortical cells matures extremely rapidly, the proportion of binocular neurons increases, and there is vulnerability to monocular deprivation.

Casaer: The few 'normal' data on glucose and oxygen consumption measured by positron emission tomography indicate that the visual occipital cortical areas become active around 10 weeks of postnatal age in infants born at term (Chugani et al 1987).

Murray: It is said that selective cell death is a way of removing neuronal misconnections. When you enucleate one eye in experimental animals and thereby decrease the amount of selective cell death in the other optic tract, is there any evidence that early, immature misconnecting axons now persist?

Blakemore: Yes. After removal of one eye early in fetal life (in cat or monkey) the entire lateral geniculate nucleus remains innervated by the one remaining eye, rather than forming left-eye and right-eye layers (Rakic 1979). Therefore, connections from the remaining eye are retained in regions of the nucleus from which they should have been displaced. The normal process of competitive displacement is accompanied by the death of ganglion cells in each eye and early removal of one eye leads to fewer ganglion cells dying in the remaining retina. Thus the competitive segregation process may partly depend on selective cell death. On the other hand, the superficially similar segregation of right-eye and left-eye geniculate fibres that occurs in the primary visual cortex after birth does not seem to involve cell death among the cells of origin of those axons in the lateral geniculate nucleus.

I can give you a very specific case in which cell death is responsible for the elimination of inappropriately connected axons and in which those aberrant connections persist if cell death is prevented. Insausti, Cowan and I studied the development of the retinal input to the midbrain visual nucleus, the superior colliculus, in the hamster (Insausti et al 1984). In the normal adult, the colliculus on each side is innervated by axons from the entire retina of the opposite eye (distributed to form a 'map' of the visual field) and also by a specific population of ganglion cells lying only in the extreme temporal retina of the ipsilateral eye, which views the middle of the visual field: these ipsilateral fibres terminate exclusively in the rostral part of the colliculus, where they lie in register with fibres from the temporal crescent of the contralateral eye, which views the same part of the visual field. In the newborn animal, however, there is a population of ipsilaterally projecting ganglion cells lying outside the temporal crescent, sparsely scattered over the whole retina, and the axons of the ipsilaterally projecting cells extend to the caudal pole of the superior colliculus. These aberrant ipsilateral ganglion cells normally die, as the terminal arborization in the colliculus adopts its adult distribution. However, if the other eye is removed at birth, the death of these aberrant cells is prevented and the ipsilateral projection persists over the whole of the colliculus.

So, one function of selective cell death is to refine projections and to eliminate misrouted axons. However, a large component of cell death is almost certainly preprogrammed, perhaps simply compensating for an initial overproduction of neurons.

Murray: To speculate about cognitive function here, is it possible that some of the hazards that we are considering (hypoxia or malnutrition) leave an *excess* of connections, and that these extra connections may produce too much 'noise' for efficient functioning of the nervous system?

Blakemore: That is an interesting possibility. There are now many examples of interhemispheric and other cortico-cortical connections (which might play a part in cognitive function) that are initially highly 'exuberant' and are refined postnatally to produce their correct distribution (e.g. Innocenti 1981, Price & Blakemore 1985). And even within an anatomically normal pattern of innervation, there is a strong likelihood that synaptic strength is being modulated, on the basis of experience, to produce invisible but functionally important changes in circuitry. The poor spatial resolution of the receptive fields of neurons in the visual cortex after deprivation is likely to be due to a failure to select the synaptic input to each cell and to the retention of input from too wide a region of the retina. One could speculate that similar refinement is taking place at much later ages in other areas of the cortex, perhaps in the temporal lobe, and that failures in this process might lead to cognitive and perceptual disorders.

Suomi: There are, in theory, at least three different ways in which an excess number of cells could be reduced. First, some cells at their initiation might be destined to be survivors and others destined to die, independent of subsequent events and experience—that is, it might be possible to identify 'survivor' cells ahead of time.

A second possibility is that the paring down is completely random, independent of any environmental contingency. The third possibility is that every cell starts off with an equal chance but its fate depends on what the cell experiences subsequently in terms of stimulation.

The question is whether the consequences of these three models differ with respect to flexibility and adaptation. Within that third possibility is the potential for generating diversity in a way that is responsive to the environmental circumstances at the time, that might well promote the organism's survival in later circumstances.

A further thought is that these different scenarios—and there are probably many more than these three—need not be universal across all developing systems. They may be biased more in one particular system, like the immune system, than in the motor reflex system, for example.

References

Atkinson J 1989 New tests of vision screening and assessment in infants and young children. In: French J, Harel S, Casaer P (eds) Child neurology and developmental disabilities. Paul H Brookes, Baltimore, p 219–229

Atkinson J, Braddick OL 1990 The developmental course of cortical processing streams in the human infant. In Blakemore C (ed) Vision: coding and efficiency. Cambridge University Press, Cambridge, p 247–253

Bitman J, Wood DL, Hamosh M, Hamosh P, Mehta NR 1983 Comparison of the lipid composition of breast milk from mothers of term and preterm infants. Am J Clin Nutr 38:300–313

Caroni P, Schwab ME 1988 Two membrane protein fractions from rat central myelin with inhibitory properties for neurite growth and fibroblast spreading. J Cell Biol 106:1281–1288

Clandinin MT, Chappell JE, Heim T 1982 Do low birth weight infants require nutrition with chain elongation-desaturation products of essential fatty acids? Prog Lipid Res 20:901–904

Chugani T, Phelps ME, Mazziotta JC 1987 Positron emission tomography study of human brain functional development. Ann Neurol 22:487–497

Dobbing J 1990 Early nutrition and later achievement. Proc Nutr Soc 49:103–118

Dobbing J, Sands J 1979 Comparative aspects of the brain growth spurt. Early Hum Dev 3:79–83

Innocenti GM 1981 Growth and reshaping of axons in the establishment of visual callosal connections. Science (Wash DC) 212:824–827

Insausti R, Blakemore C, Cowan WM 1984 Ganglion cell death during development of ipsilateral retino-collicular projection in golden hamster. Nature (Lond) 308:362–365

Kiorpes L, Boothe RG, Carlson MR, Alfi D 1985 Frequency of naturally occurring strabismus in monkeys. J Pediatr Ophthalmol Strabismus 22:60–64

Neuringer M, Connor WE 1986 n3 fatty acids in brain and retina: evidence for their essentiality. Nutr Rev 44:285–294

Price D, Blakemore C 1985 Regressive events in the postnatal development of association projections in the visual cortex. Nature (Lond) 316:721–724

Rakic P 1979 Genesis of visual connections in the rhesus monkey. In: Freeman RD (ed) Developmental neurobiology of vision. Plenum Publishing Corporation, New York, p 249–260

Sur M, Pallas L, Roe AW 1990 Cross-modal plasticity in cortical development: differentiation and specification of sensory neocortex. Trends Neurosci 13:227–233

Fetal brain development and later schizophrenia

Robin M. Murray, Peter Jones and Eadbhard O'Callaghan

Genetics Section, Department of Psychological Medicine, King's College Hospital and the Institute of Psychiatry, de Crespigny Park, London SE5 8AF, UK

Abstract. Computed tomography and magnetic resonance imaging studies have shown cerebral ventricular enlargement and a decreased volume of temporal lobe structures in a proportion of schizophrenic patients. Neuropathological investigations confirm these findings and also show diminished volume of the hippocampus and abnormal pre-alpha cell clusters in the parahippocampal gyrus. Compared with controls, schizophrenic patients are more likely to have minor physical anomalies, to have a history of obstetric complications, and to have been born in the late winter. Together the evidence regarding structural brain abnormalities and epidemiology suggests that a significant proportion of cases of schizophrenia have their origins in fetal or neonatal life. The mechanisms involved in the aberrant neurodevelopment remain obscure but some impairment of neuronal migration is an appealing hypothesis.

1991 The childhood environment and adult disease. Wiley, Chichester (Ciba Foundation Symposium 156) p 155–170

Schizophrenia is usually diagnosed by the presence of characteristic delusions and hallucinations. However, in addition to these florid or positive symptoms, many patients also demonstrate a negative syndrome comprising symptoms such as social withdrawal, lack of motivation and drive, together with subtle cognitive defects and neurological 'soft' signs such as abnormalities of coordination and tone. Although less conspicuous than the positive syndrome, the negative syndrome carries a worse prognosis (Crow 1985).

Delusions and hallucinations seldom appear before adolescence, but it has become clear that a proportion of patients exhibit abnormalities of intellect and personality in childhood long before the positive symptoms appear. Thus, Aylward et al (1984) concluded that premorbid IQ scores of schizophrenics obtained during childhood, adolescence and early adult life are lower than the scores of their siblings and peers from similar social class backgrounds. Similarly, findings regarding premorbid personality from (a) follow-up studies of attenders at child guidance clinics, (b) follow-back studies, which examine school reports, and (c) prospective high risk studies, demonstrate that approximately 25–50%

of schizophrenics showed abnormal behaviour in childhood. We (A. Foerster et al, unpublished) have demonstrated that compared to those children who later develop affective psychosis, preschizophrenics are more likely to exhibit schizotypal and schizoid traits (i.e. unusual patterns of thought and communication) as well as poor social adjustment as early as ages 5–11 years. We have suggested elsewhere that cognitive and personality abnormalities in childhood are, in fact, the forerunners of the negative syndrome of adult life, and that this is why they augur a poor outcome (Murray et al 1988).

Neuroimaging studies

A proportion of schizophrenic patients have enlarged cerebral ventricles on computed tomography (CT) scan. These are not, as was initially thought, a consequence of some neurodegenerative process, but are present at the onset of illness and do not progress over almost a decade. They have been postulated to be related to the negative rather than the positive syndrome, are more pronounced in males than females, and appear to be associated with poor adjustment in childhood (Shelton & Weinberger 1986). Reveley et al (1982) showed that the schizophrenic members of monozygotic twins discordant for the condition have significantly larger ventricles than their non-schizophrenic co-twins. As cerebral ventricular size is normally highly correlated in monozygotic twins, the cause of the enlargement must be environmental.

Magnetic resonance imaging (MRI) studies have confirmed the finding of enlarged ventricles on CT and have shown a smaller volume of brain structures, particularly in the temporal lobes. Suddath et al (1990) examined 15 pairs of monozygotic twins discordant for schizophrenia. The schizophrenics showed enlarged lateral and third ventricles and smaller hippocampi than their well co-twins. In 13 out of 15 cases, the total volume of grey matter in the left temporal lobe was smaller in the schizophrenic twin than in the normal co-twin, a finding consistent with other evidence implicating left-sided abnormalities in schizophrenia.

There has been considerable dispute over the clinical correlates of CT abnormalities. However, in an interesting paper, Johnstone et al (1989) demonstrated that cognitive impairment was correlated with decreased brain area in early-onset, but not late-onset cases. This goes along with much other evidence suggesting that early onset cases of schizophrenia may be different from late onset cases; for example, the former are more likely than the latter to be male and to have shown personality and cognitive deficits in childhood (Aylward 1984, A. Foerster et al, unpublished). Together, these facts remind one that childhood neurodevelopmental disorders such as autism and dyslexia are all much more common in males.

Early developmental factors

Schizophrenics are more likely to have a history of obstetric complications than other psychiatric patients, their siblings, and normal controls (Murray et al 1988). Most of these studies have relied on retrospective recall by the mothers of patients, but the findings have been replicated using contemporaneous birth records; indeed, O'Callaghan et al (1990) have shown a high correlation between mothers' recall and original records. The term 'obstetric complications' is used here in the broad sense, to cover any detectable deviation from normal during pregnancy, birth and early neonatal life. Obstetric complications appear particularly in the histories of male schizophrenics, are associated with early onset, and in many, but not all studies predict increased ventricular size (Lewis et al 1989).

A further indicator of developmental disruption comes from the excess of minor physical anomalies (MPAs) (for example, malformed ears, steepled palate) that are found in schizophrenia. These trivial abnormalities in ectodermal development are of little consequence in themselves. However, they appear to be markers of developmental abnormality in the central nervous system which, of course, derives from the ectoderm, and are known to occur in excess in patients with developmental disorders. Three studies have found an excess of MPAs among schizophrenic patients when compared to controls (see Green et al 1987). Once again, MPAs are more common in male schizophrenics, and predict a younger age of onset, and possibly also cognitive impairment (Green et al 1987, Waddington et al 1990). The work of C. Mellor (personal communication) showing that schizophrenics have more dermatoglyphic abnormalities than controls further buttresses the evidence that predisposition to schizophrenia in some individuals is determined prenatally.

Schizophrenics show a slight increase (7–15%) in births in the late winter and spring. This excess tends to be noted more frequently in those without a family history of the disorder (Shur 1982, Sacchetti et al 1989; E. O'Callaghan et al unpublished paper, 5th Biennial Winter Workshop on Schizophrenia, 28 Jan–4 Feb 1990). Furthermore, among familial cases, ventricular size does not vary according to season of birth, but it does among non-familial cases (Sacchetti et al 1987, Jones et al 1991). These data have been taken to reflect some seasonal factor impairing fetal brain development, causing enlargement of cerebral ventricles, and predisposing to later schizophrenia.

There is some, not yet conclusive, evidence that the incidence and severity of schizophrenia may be declining in certain Western countries (Der et al 1990). We (Murray et al 1990) have pointed out that in England and Wales the decline began some two decades after a similar fall in early neonatal mortality; we speculate that improved fetal health due to factors such as better maternal nutrition or decreased infectious disease may have diminished the number of neonates born at risk of later schizophrenia.

Neuropathology and neurodevelopment

Cell proliferation

For most of this century, neuropathological investigations of schizophrenia have produced contradictory results, but in the last seven years a consensus has begun to emerge. The brains of schizophrenics show a decrease in weight, estimated at 6% by Brown et al (1986). Bogerts and colleagues (1985) showed a greater decrease (20–30%) in the volume of the limbic temporal lobe (amygdala, hippocampus, parahippocampal gyrus), a finding which is consistent with the evidence of particular enlargement of the temporal horn of the lateral ventricles (Colter et al 1987).

Two groups (Falkai & Bogerts 1986, Jeste & Lohr 1989) found decreased cell counts in the hippocampus. Gliosis is the usual glial cell reaction to neuronal inflammation or damage, and its absence in most schizophrenics (Falkai & Bogerts 1986) suggests that the decreased cell counts may have arisen through some failure of neuronal proliferation.

The hippocampus has received considerable attention from developmental neurobiologists, because it has a relatively well-defined structure which can be followed throughout neurogenesis. It develops exclusively from the ventricular zone (Nowakowski & Rakic 1981) and reaches half its adult volume by the time of birth, and its adult size by approximately two years of age. The smaller volume of, and decreased cell numbers in, the hippocampi of schizophrenics could therefore be a consequence of impaired proliferation in the ventricular zone.

Cell migration

The basic structure of the hippocampus is formed by young neurons migrating from the ventricular surface, a process in which they are guided by the radial glia. If neuronal migration is disrupted in the fetus, abnormally positioned cells result. Kovelman & Scheibel (1984) have reported disarray of the normally regimented ranks of pyramidal neurons in the CA1/CA2 regions of the hippocampus in schizophrenia. A subsequent study from the same group failed to find a statistical difference in neuronal disorganization between schizophrenic subjects and controls but did report greater disarray among those schizophrenics with a more severe psychosis.

Jakob & Beckmann (1986) noted that in the brains of some adult schizophrenics, particularly those with early onset, groups of pre-alpha neurons, normally located in the superficial layers of the entorhinal cortex, are displaced deep to their expected position. Falkai et al (1988) replicated this finding by calculating the distance between the pial surface of the entorhinal cortex and the centre of the pre-alpha cell clusters, and showing this to be relatively increased in schizophrenia. Some defect in the control of embryonic cell migration is an appealing explanation.

Cell death

It is well known that a large proportion of all cells generated in the developing nervous system die by the time it is mature. This process of selective neuronal death appears to eliminate early errors of connection, and several authors have suggested that this might be abnormal in schizophrenia. Benes et al (1986) showed neuronal density to be lower in schizophrenics in layer 6 of the prefrontal cortex, layer 5 of the cingulate gyrus and layer 3 of the motor cortex. These authors suggest that their findings could have arisen from 'an accelerated process of neuronal drop out early in life, perhaps related to a perinatal insult'. Such a process could result in a compensatory decrease in selective cell death, and the persistence of immature patterns of cells and their connections into adult life. Indeed, Benes et al (1987) later reported increased numbers of vertical axons in the cingulate gyrus of schizophrenics. Similarly, Deakin et al (1989) have attributed the apparent abnormally dense glutamatergic innervation of the orbital frontal cortex in schizophrenia to 'an arrest or failure of the process by which transient callosal connections normally are eliminated during development'.

Goodman (1989) believes that such anomalous patterns of neuronal connections could give rise to some of the misconceptions so typical of schizophrenic thought. Malfunctions of neuronal networks within the septo-hippocampal system have been particularly implicated by Gray et al (1991), who consider that this system monitors incoming perceptual information, and compares it with information stored in memory relating to previous stimuli. These authors hypothesize that dysfunction of this 'comparator' causes a mismatch between the internal model of, and the actual state of, the real world, which produces the positive symptoms of schizophrenia.

Such theories do not explain why the positive symptoms do not appear until late adolescence and early adult life. The most likely explanation is that the cytoarchitectonic abnormalities remain largely dormant until some maturational process brings them to light. Myelination of axons is one possibility; Benes (1989) has shown strikingly increased myelination of the subicular and presubicular regions during late adolescence. She notes that these structures have a strategic location within the cortical limbic circuitry of the brain, and asks whether myelination of one or two linkages in this circuitry could in some way permit a pre-existing but latent defect in the hippocampus to become manifest.

Aetiology

What causes the developmental abnormalities found in the brains of some schizophrenics? There is ample evidence of a major genetic contribution to aetiology from family, twin and adoption studies. It appears that at least two-thirds of the variance in liability to schizophrenia is genetic, but the mode of transmission is unclear. Single-gene, polygenic and heterogeneity theories have all been proposed, but existing statistical methods are unable to distinguish between the adequacy of these (Murray et al 1986).

Until now, genetic research into schizophrenia has been preoccupied with the adult clinical phenotype and has continued largely oblivious of the neuroimaging and neuropathological studies that have just been described. However, it is now clear that psychiatric geneticists will have to turn their attention towards understanding the genetics of brain development. Our knowledge of this is very primitive but we do have some information about the genetic control of hippocampal development, because its cytoarchitecture in mice is predictably affected by four well-characterized single-gene defects (Nowakowski 1987). In the NZB/BINJ mutation, abnormally positioned neurons have migrated too far, while in the HLD, Dreher and Reeler mice, the abnormally positioned cells have not migrated far enough. Conrad & Scheibel (1987) were struck by the similarities in the pattern of lamina organization produced by these defects and that seen in schizophrenia, and suggest that they may 'serve as a conceptual model or perhaps as a caricature of a more subtle developmental anomaly in the schizophrenias'.

It is obviously over-simplistic to regard the genes underlying these murine mutants as 'candidate genes' for human schizophrenia, but we may derive some clues from the morphological consequences of these defects. For example, very similar neuronal phenotypes can be produced by environmental hazards such as ionizing radiation, mercury poisoning and the fetal alcohol syndrome, providing these operate within the crucial period of neuronal migration in the mouse (Nowakowski 1987).

Is it possible that in humans, as in the mouse, a similar hippocampal phenotype (and therefore clinical picture) could be produced by mutant genes and early environmental hazards? As noted earlier, a minority of schizophrenics have a history of pre- or perinatal complications. Ischaemia affecting the fetus (chronically) or neonate (acutely) has been repeatedly proposed as a possible mediator. Could this be related to the fact that the pyramidal cells in the CA1 region of the hippocampus are among the most vulnerable in the human brain to mild ischaemia? Similarly, the winter excess of schizophrenic births has been attributed to viral infection, and maternal influenza in the second trimester of pregnancy is reported to increase the risk to the fetus of schizophrenia in later life (Mednick et al 1988). Could this be related to claims that neuraminidase-containing viruses such as influenza virus can interfere with intercellular adhesiveness, and possibly perturb the migration of hippocampal neurons (Nowakowski 1987, Conrad & Scheibel 1987)?

Among schizophrenics, a history of an affected relative and a history of obstetric complications tend to segregate separately. Also, as noted earlier, it is non-familial schizophrenics who show an excess of winter births. These facts suggest that some early cerebral insult may induce phenocopies in those not obviously predisposed genetically. However, among the children of schizophrenics, those who suffer perinatal complications have the highest risk of later developing schizophrenia themselves; this implies that the two types of causal factors may act additively, perhaps on the same system. It may be that some cases of schizophrenia result from single-gene defects, while a proportion may be wholly environmental in origin; others may inherit a pattern of neuronal development especially vulnerable to environmental perturbations (Murray et al 1988).

It will be evident from the above that in seeking the cause of schizophrenia, we should not expect to find a gene or genes that code directly for delusions or hallucinations. Instead, the control of brain development may be defective and produce some subtle form of 'faulty neuronal wiring'. The process of neuro-development is so wonderfully complex that a variety of mutations affecting glia, or the proliferation or migration of neurons, could all produce a similar phenotype. Indeed, there is growing evidence that an identical neuropathological and clinical picture can result from environmental interference with the same process.

Acknowledgements

We are grateful for the support of the Medical Research Council, the Wellcome Trust and the Leverhulme Trust, and for the advice of Drs S. W. Lewis, R. Kerwin and P. MacGuire.

References

Aylward E, Walker E, Bettes B 1984 Intelligence in schizophrenia. Schizophr Bull 10:430–459

Benes FM 1989 Myelination of cortical–hippocampal relays during late adolescence. Schizophr Bull 15:585–593

Benes FM, Davison J, Bird ED 1986 Quantitative cyto-architectural studies of the cerebral cortex in schizophrenics. Arch Gen Psychiatry 43:31–35

Benes FM, Majocha R, Bird ED, Marotta CA 1987 Increased vertical axon counts in cingulate cortex of schizophrenics. Arch Gen Psychiatry 44:1011–1021

Bogerts B, Meertz E, Schonfeldt-Bausch R 1985 Basal ganglia and limbic system pathology in schizophrenia: a morphometric study of brain volume and shrinkage. Arch Gen Psychiatry 42:784–791

Brown R, Coulter M, Corsellis JAN et al 1986 Postmortem evidence of striatal brain changes in schizophrenia. Arch Gen Psychiatry 43:36–42

Colter N, Bruton CJ, Johnstone EC, Roberts GW, Brown R, Crow TJ 1987 Neuro-pathology of schizophrenia. II. Neuropathol Appl Neurobiol 13:499–500

Conrad AJ, Scheibel AB 1987 Schizophrenia and the hippocampus: the embryological hypothesis extended. Schizophr Bull 13:577–587

Crow TJ 1985 The two syndrome concept: origins and current status. Schizophr Bull 11:471–485

Deakin JFW, Slater P, Simpson MDC et al 1989 Frontal cortical and left temporal glutamatergic dysfunction in schizophrenia. J Neurochem 52:1781–1786

Der G, Gupta S, Murray RM 1990 Is schizophrenia disappearing? Lancet 335:513–516

Falkai P, Bogerts B 1986 Cell loss in the hippocampus of schizophrenics. Eur Arch Psychiatry Neurol Sci 236:154–161

Falkai P, Bogerts B, Rozumek M 1988 Limbic pathology in schizophrenia: the entorhinal region—a morphometric study. Biol Psychiatry 24:518–521

Goodman RN 1989 Neuronal misconnections and psychiatric disorder. Is there a link? Br J Psychiatry 154:292–299

Gray JA, Feldon J, Rawlins JWP, Hemsley DR, Smith AD 1991 The neuropsychology of schizophrenia. Behav Brain Sci, in press

Green MF, Satz P, Soper HV, Kharabi F 1987 Relationship between physical anomalies and age of onset of schizophrenia. Am J Psychiatry 144:666–667

Jakob H, Beckmann H 1986 Prenatal developmental disturbances in the limbic allocortex in schizophrenics. J Neural Transm 65:303–326

Jeste DV, Lohr JB 1989 Hippocampal pathologic findings in schizophrenia: a morphometric study. Arch Gen Psychiatry 46:1019–1024

Johnstone EC, Owens DG, Bydder GM, Colter N, Crow TJ, Frith CD 1989 The spectrum of structural brain changes in schizophrenia: age of onset as a predictor of cognitive and clinical impairment and their cerebral correlates. Psychol Med 19:91–103

Jones P, Owen MJ, Goodman R, Lewis SW, Murray RM 1991 Neurodevelopment and the chronological curiosities of schizophrenia. In: Proceedings of NATO Conference on Fetal Neurodevelopment in Schizophrenia. Plenum Press, New York, in press

Kovelman JA, Scheibel AB 1984 A neurohistological correlate of schizophrenia. Biol Psychiatry 19:1601–1621

Lewis SW, Murray RM, Owen MJ 1989 Obstetric complications in schizophrenia: methodology and mechanisms. In: Schultz SC, Tamminga CA (eds) Schizophrenia: scientific progress. Oxford University Press, New York, p 56–69

Mednick SA, Machon RA, Huttenen MO, Bonnett D 1988 Adult schizophrenia following prenatal exposure to an influenza epidemic. Arch Gen Psychiatry 45:18–192

Murray RM, Reveley AM, McGuffin P 1986 Genetic vulnerability to schizophrenia. In: Roy A (ed) Schizophrenia. Saunders, Philadelphia (Psychiatric Clinics of North America) p 3–16

Murray RM, Lewis SW, Owen MJ, Foerster A 1988 The neurodevelopmental origins of dementia praecox. In: McGuffin P, Bebbington P (eds) Schizophrenia: the major issues. Heinemann, London, p 90–107

Murray RM, Gupta S, Der G 1990 Trends in schizophrenia. Lancet 335:1214

Nowakowski RS 1987 Basic concepts of CNS development. Child Dev 58:568–595

Nowakowski RS, Rakic P 1981 The site of origin and route and rate of migration of neurones to the hippocampal region of the Rhesus monkey. J Comp Neurol 196:129–154

O'Callaghan E, Larkin C, Waddington JL 1990 Obstetric complications in schizophrenia: validity of maternal recall. Psychol Med 20:89–94

Reveley AM, Reveley MA, Clifford CA, Murray RM 1982 Cerebral ventricular size in twins discordant for schizophrenia. Lancet 1:540–541

Sacchetti E, Vita A, Battaglia A et al 1987 Season of birth and cerebral ventricular enlargement in schizophrenia. In: Cazzullo CL, Invernizzi G, Sacchetti E, Vita A (eds) Etiopathogenetic hypothesis of schizophrenia. MTP Press, Lancaster

Sacchetti E, Vita A, Giobbio GM, Dieci M, Cazzullo CL 1989 Risk factors in schizophrenia. Br J Psychiatry 155:266–267

Shelton RN, Weinberger DR 1986 CT studies in schizophrenia. A review and synthesis. In: Nasrallah H, Weinberger DR (eds) The neurology of schizophrenia. Elsevier, Amsterdam

Shur E 1982 Season of birth in high and low genetic risk schizophrenics. Br J Psychiatry 140:410–415

Suddath RL, Christison GW, Torry EF 1990 Anatomical abnormalities in the brains
of monozygotic twins discordant for schizophrenia. N Engl J Med 322:789–794
Waddington JL, O'Callaghan E, Larkin C 1990 Physical anomalies and neurodevelopmental abnormality in schizophrenia: new clinical correlates. Schizophr Res
3:90

DISCUSSION

Parnas: We have completed a 27-year follow-up of children of schizophrenic
mothers, and some of our preliminary findings support what Professor Murray
was saying.

The study began in Copenhagen 1962 with an examination of 207 children
of schizophrenic mothers and 104 control children. A variety of premorbid data
were collected, including obstetric data. (In Denmark, each delivery is required
by law to be assisted by a midwife, who completes a special report; a national
register contains all the obstetric data so obtained.) The children have been
followed up regularly until a final diagnostic follow-up in 1989. Our preliminary
data analysis indicates some results that are relevant to the topic of this
symposium.

First, we showed that chronic schizophrenic subjects had many more obstetric
complications than any of the other diagnostic groups. What is especially
interesting is that chronic schizophrenics differ both from schizophrenics who
remit and from schizotypal subjects—that is, people with attenuated features
of the negative syndrome to which Professor Murray referred. So an early onset,
chronic schizophrenic outcome is definitely associated with perinatal
complications.

Our preliminary findings also suggest that chronic schizophrenics have
enlarged cerebral ventricles, as Professor Murray was saying. It is interesting
to note that adult ventricular enlargement is predicted by low birth weight and
by perinatal complications. What is even more interesting is that we find an
interaction between genetic predisposition and obstetric complications in the
prediction of enlargement of the ventricles. The interaction with genetic
predisposition in our sample is indexed by paternal diagnosis; that is, all 207
children had a schizophrenic mother, but if in addition their fathers suffered
from a schizophrenia spectrum diagnosis, the association between perinatal
complications and adult brain atrophy is much stronger (Cannon et al 1989).

Barker: Could Professor Murray sharpen up the description of the obstetric
complications? What do these complications include?

Murray: Essentially, obstetric complication scales have been developed which
are non-specific and measure everything from toxaemia in the mother to length
of labour. Some people suggest that greater length of labour, and other
complications particularly likely to be associated with hypoxia, might especially

predispose to later schizophrenia. Such a line of reasoning suggests that acute or chronic hypoxia may cause intraventricular bleeding with the ventricular enlargement being consequent upon that. However, there is evidence that events earlier on in pregnancy (e.g. maternal viral infection) may be most important. It could be that longer labour, or low birth weight, might be consequent on some disorder of development in the second trimester. Furthermore, if neuronal migration is abnormal in schizophrenia, as many suggest, then that's too early to be a consequence of hypoxia at birth.

Barker: Children with Down's syndrome tend to be born prematurely and tend to be small for dates, which is consistent with your view that the damage in schizophrenia could occur long before delivery.

Murray: It has been argued that obstetric complications may be an epiphenomenon of the genetic predisposition to schizophrenia, but I would dispute this. In Joseph Parnas's study, all the mothers had schizophrenia by definition, so the offspring were a genetically predisposed group, and the obstetric complications compounded that. However, among schizophrenics as a whole, those with a family history of psychosis and those with a history of obstetric complications tend to separate out.

Barker: The question is whether obstetric complications are an epiphenomenon of an environmental insult in the second or third trimester of pregnancy.

Murray: We don't know. There is one study from Helsinki, where Mednick and colleagues (Mednick et al 1988) looked at the Asian flu epidemic which reached Helsinki in October 1957. They looked through the mental hospitals in the 1970s and 1980s and claimed that babies who were in the second trimester when their mothers were potentially exposed to the influenza epidemic had a 2–3 times increased risk of subsequent schizophrenia. Kendell & Kemp (1989) have criticized the statistics of that study, and have tried to replicate it in Scotland, but have not done so convincingly.

Richards: May I make some comments on the birth complications? One of the problems is an historical one, that a long labour in 1948 is very different from a long labour in 1970. Over the period when most of these studies were done there have been dramatic changes in the nature of obstetrics and so in the nature of birth.

You discussed seasonal effects: I wonder if those are accounted for by low birth weight, which shows a seasonal effect of a similar kind.

As to genetic effects, genetics has to work through a developmental process, and perhaps sometimes it works through the mothers in pregnancy and in the ways in which they may influence the growing fetus. Perhaps it is not a gene that causes the hole in the head, as it were, but a gene that does something to the mother's uterus or another aspect of her physiology which in turn leads to changes in fetal development. These fetal changes might, in turn, lead on to situations that may be called 'low birth weight' or 'birth complications'. Or there may be various inputs into that pathway that could come from different events in different cases. Perhaps we should not be seeking a single pathway or a single cause.

Murray: That is very interesting. Psychiatrists keep trying to see if there is a gene which directly produces the clinical phenomena of schizophrenia; yet it is ridiculous to suppose there is a gene for hearing a particular kind of voice, as in schizophrenic hallucinations. As you say, there may be a gene, or genes, which is, or are, involved in fetal development; or it could be a maternal effect. One study showed that ventricular size is greater in schizophrenics with a schizophrenic mother than in those with a schizophrenic father. Could that be the result of some maternal, intrauterine effect?

Richards: There are many papers which suggest connections between birth complications and other rather imprecise factors, such as difficulties in getting pregnant, or menstrual irregularities. But we have little knowledge of processes that might link such events.

Barker: In the history of research on Down's syndrome there were many papers on its strong relation with obstetric complications. Less was heard about that after the identification of the chromosomal abnormality.

Meade: Could I ask Professor Murray to comment on the work by George Brown on life events (Brown et al 1986), as a kind of environmental interaction with the aspects you have been talking about? My understanding is that these events probably relate more to the relapse of schizophrenia than to its onset. Exposure to adverse life events, like being thrown out of your home or out of hospital, or having a row with someone, is associated with a tendency to relapse. Is that due entirely to the event, or is it in any way an interaction with the factors you have been discussing?

Murray: 'Life events' in depression are indeed thought to be causal; adverse life events over a period of months may induce depression in someone not otherwise particularly predisposed. In schizophrenia it's thought to be the case that adverse life events precipitate or bring forward the schizophrenic symptoms which would have occurred anyway, but at a later date. We looked at a series of one hundred psychotics in whom we have data on life events, and also have data on psychiatric disorder in their relatives. We wondered if there could be (a) a biological type of schizophrenia, with a genetic predisposition or developmental difficulty, and (b) a social environmental type of schizophrenia. We looked at life events versus biological predisposition (unpublished). It turned out that adverse life events were clustered in those individuals who had a relative affected by psychosis. It was as if individuals might inherit a pattern of neuronal development, or of neurotransmission, which made them more susceptible to break down after some adversity by developing the psychotic symptom. On the other hand, those schizophrenics who had evidence of early 'environmental' damage had not experienced an excess of adverse life events: that is, those who were of low birth weight or had suffered obstetric difficulty seemed to be different, in that schizophrenia wasn't precipitated in them by adverse life events.

Meade: That is very interesting, because it brings the life event into the onset of the disease, not just its relapse, in genetically disposed individuals.

Murray: Yes; that would be accepted. But I don't think many researchers would think that normal people could be precipitated into schizophrenia by adverse life events, though they might be precipitated into depression by such events.

Golding: On the obstetric complications, this sounds very like the story that was current about cerebral palsy, where it was assumed that this was a birth injury and obstetric complications were quoted. The bulk of the evidence now is that such children are already damaged before the mother goes into labour, and all the apparent associations with obstetric complications are due to the damaged child who isn't reacting appropriately to delivery and appears asphyxiated at birth (Nelson 1988). It sounds as though your evidence supports some event happening earlier in pregnancy, but again, from what published data I have seen, it's difficult to untangle what might be there.

The temporal variation in schizophrenia that you have indicated is unlike that found with cerebral palsy, where there is no sudden decrease in the incidence with a falling infant mortality rate (Emond et al 1989). I have been involved in a study following up the 1958 birth cohort, identifying schizophrenic subjects. A scoring system weighting the factors predictive of perinatal mortality was not predictive of schizophrenia.

Murray: There is so much evidence that obstetric complications *are* associated with later schizophrenia that I don't think they can be discounted, but perhaps that study might imply, along with Joseph Parnas's data, that individuals might need to have the mutant gene(s) and then have the early environmental difficulty to cause schizophrenia. I would accept the point that what we call obstetric difficulties are largely *perinatal* difficulties; we can detect them, but we can't detect what's been going on at four, five or six months of pregnancy; the perinatal difficulties may be a consequence of earlier undetectable adversity.

Golding: It's difficult to tie that in with your temporal variation, where you have an apparent drop in incidence, which appears to mirror the perinatal mortality, and yet when you look at it you cannot see any reason for that.

Murray: The argument continues about whether there is a real fall in the incidence of schizophrenia, or merely in the number of schizophrenics hospitalized. I believe there is a real fall. But, to me, the neonatal mortality rate is not an indicator of something going wrong at birth, but an indicator of malnutrition or some earlier developmental difficulty in the fetus which puts it more at risk of death at the time of birth.

Parnas: I would like to emphasize this perinatal component, which is underestimated. For instance, your own study (Reveley et al 1982) on identical twins showed that the affected twin has ventricular enlargement whereas the non-affected twin has not, which strongly suggests that the perinatal insult is not an epiphenomenon of genetic loading and is perhaps not the consequence of something occurring very early during pregnancy. A Swiss study from the University of Lausanne has just been completed with excellent perinatal data; this study shows a clear-cut association between signs of asphyxia at birth and later schizophrenia (P. Bovet, personal communication). In this study the index of perinatal complications was not simply a vague indicator of something going wrong; they could measure the signs of asphyxia and could also partial out the seasonality effects. Even though these were partialled out, perinatal complications were still important as predictors of adult schizophrenia.

Hanson: We have heard reference to 'schizotypes'. Have the criteria for the diagnosis of schizophrenia remained constant since the war? Is it possible that the decrease in incidence that you see from 1960 is due to a sharpening up of the 'true' schizophrenics which have been diagnosed?

Murray: That is one of the issues—whether the concept of schizophrenia has narrowed over this period. If that were the case, you might expect to see a compensatory rise in some other diagnostic entity—say, manic-depressive psychosis. We haven't found that. One can't do much more on the national data because one cannot obtain case records of all the subjects. Therefore, we are now going back to our local Camberwell data and applying five operational definitions of schizophrenia to the case notes of all possibly schizophrenic individuals seen over 20 years. We shall then be able to work out the incidence rate for schizophrenia between 1964 and 1984, using the same criteria over the whole period. However, we can't exclude this possibility of a change in diagnostic habits yet.

Hanson: Another question is whether the treatment, or non-treatment, of such people has changed over that time. Because, if schizophrenia is not an all-or-nothing phenomenon, but a spectrum with some people who may be latent schizophrenics, it might be that a certain treatment strategy would bring the condition out, and that could explain a change in the incidence.

Murray: That is conceivable. Or could it be, for example, that nowadays, every time somebody behaves a bit oddly and visits their general practitioner, they are given an antipsychotic drug and therefore never reach a psychiatrist? Against that, you might expect that the rise in drug abuse, particularly drugs such as ecstasy or amphetamines or LSD, which are known to precipitate schizophrenic symptoms, might have increased the frequency with which schizotypal personality has switched into frank schizophrenia.

Lucas: Could I sharpen up the suggested relationship between prematurity and schizophrenia? A mild degree of prematurity goes with congenital syndromes but extreme prematurity, with gestation of less than 32 weeks, which has been increasingly compatible with survival since the early 1970s, is usually not associated with congenital syndromes. These are essentially normal babies that are born preterm, so far as we can see. I am interested to know whether there is a correlation between the degree of prematurity and schizophrenia. After all, prematurity is an important cause of brain damage and of large cerebral ventricles and other pathology that you see in schizophrenia. And if there were a relationship between extreme prematurity and schizophrenia, the incidence ought to be going up again now. We are now about 20 years on from the point where these very premature babies were beginning to survive.

Murray: Unfortunately, there are no data available concerning the extent of prematurity. I agree that if extreme prematurity is associated with the type of neurodevelopmental damage which might put individuals at risk of schizophrenia, then at some point we would expect to see the rates going up again.

I should mention that there's also an argument concerning whether schizophrenia is a 'recent' disease. There is an opinion that schizophrenia became much more common in the mid-19th century, and that this is why so many asylums were built in the latter part of that century. Some people believe that this increase was a consequence of early industrialization, when the rural population moved to an inner city environment. We know that the present-day children of Afro-Caribbean immigrants to Britain seem to have very high rates of schizophrenia, up to 16 times higher than in the general UK population. Could they in some way be having the same experience as rural English people moving to large cities in the 19th century?

Rutter: The association between obstetric complications and schizophrenia seems solid enough, but there must be a good deal of doubt about its meaning (see Goodman 1988). The complications constitute a most diverse collection without any discernible pattern; in particular, they do not markedly involve those that carry the highest risk of brain damage. Moreover, if they did index prenatal or perinatal damage, it might be expected that the time trend for schizophrenia should parallel that found for cerebral palsy (Hagberg et al 1989). But, as you have shown, it does not.

Also, it is necessary to recognize that the use of 'non-familiality' as an index of non-genetic cases is highly hazardous, particularly with a low frequency disorder, as Eaves et al (1986) pointed out. How sure are you that there are non-genetic varieties of schizophrenia? Admittedly, the data are limited, but, for example, the offspring of non-schizophrenic co-twins studied by Gottesman & Bertelsen (1989) had the same risk for schizophrenia as the offspring of the schizophrenic twins. If that finding holds up, it would suggest that there are not separate genetic and non-genetic varieties of schizophrenia.

Murray: That study by Gottesman & Bertelsen (1989) was based on very small numbers, so one cannot accept the findings without replication. You have also to think about what kind of schizophrenics successfully reproduce. In our most recent study, out of 56 DSM III schizophrenics, 52 had no children. Schizophrenic individuals are so handicapped that they very rarely have children. So it seems to me that Gottesman & Bertelsen are looking at a type of schizophrenia which may be much milder than the type that I am looking at.

Another issue is that of aetiological heterogeneity. I think that obstetric or early environmental factors may account for 15–20% of what we call schizophrenia, but also that a genetic predisposition leading to impaired neurodevelopment will only account for a portion of schizophrenia. Perhaps schizophrenia will prove to be like mental handicap, with a whole range of different syndromes.

The point you make regarding the crudeness of the familial/sporadic split is often made. However, if instead you measure morbid risk in the relatives of schizophrenics who have had obstetric complications, the morbid risk of schizophrenia appears lower in these relatives than in the relatives of those schizophrenics without obstetric complications (Foerster et al 1991).

A final point relates to sex differences: why is schizophrenia more common in males, and why should it be more severe? If you take males with schizophrenia and look at their relatives, those relatives have a lower risk of schizophrenia than the relatives of female schizophrenics (Goldstein et al 1990). It is almost as if there is a greater genetic predisposition in female than male schizophrenia. Males are, of course, more prone to environmental neurodevelopmental insult, and this may explain the excess of schizophrenia which is found in young males.

Mott: Surely the key must be some biochemical correlate of the morphological changes, and in trying to sort out the aetiology, knowing something about the biochemical abnormalities will be vital. What is known about the biochemical abnormalities that exist in schizophrenic brains, particularly in the areas affected morphologically?

Murray: The classic theory is the dopamine theory; if you give drugs that release dopamine, schizophrenic symptoms are worsened, whereas dopamine blockers make the symptoms better. However, there is no significant dopaminergic innervation of the hippocampal and parahippocampal regions. The biochemical abnormalities that have been shown in these areas are glutamatergic: hypoxic damage to neonatal rats can induce lasting changes in glutamate receptors in the hippocampal region, and there are abnormalities of glutamate in the hippocampal region in schizophrenic patients (Kerwin et al 1991).

References

Brown GW, Harris TO, Bifulco A 1986 Long-term effects of early loss of parent. In: Rutter M, Izard CE, Read PB (eds) Depression in young people: developmental and clinical perspectives. Guildford Press, New York, p 251–296

Cannon TD, Mednick SA, Parnas J 1989 Genetic and perinatal determinants of structural brain deficits in schizophrenia. Arch Gen Psychiatry 46:883–889

Eaves LJ, Kendler KS, Schulz SC 1986 The familial-sporadic classification: its power for the resolution of genetic and environmental etiologic factors. J Psychiatr Res 20:115–130

Emond A, Golding J, Peckham C 1989 Cerebral palsy in two national cohort studies. Arch Dis Child 64:848–852

Foerster A, Lewis S, Owen M, Murray RM 1991 Do risk factors for schizophrenia also predict poor premorbid functioning in psychosis? Schizoph Res, in press

Goldstein JM, Faraone SV, Chen WJ, Tolomiczencko GS, Tsuang MT 1990 Sex differences in the familial transmission of schizophrenia. Br J Psychiatry 156:819–826

Goodman R 1988 Are complications of pregnancy and birth causes of schizophrenia? Dev Med Child Neurol 30:391–395

Gottesman II, Bertelsen A 1989 Confirming unexpressed genotypes for schizophrenia. Risks in the offspring of Fischer's Danish identical and fraternal discordant twins. Arch Gen Psychiatry 46:867–872

Hagberg B, Hagberg G, Olow I, Wendt ZV 1989 The changing panorama of cerebral palsy in Sweden. V. The birth years 1979–82. Acta Paediatr Scand 78:283–290

Kendell RE, Kemp IW 1989 Maternal influenza in the etiology of schizophrenia. Arch Gen Psychiatry 46:878–882

Kerwin R, Patel S, Meldrum B 1991 Quantitative autoradiographic analysis of the glutamate receptor system in the hippocampal formation in normal and schizophrenic brain post-mortem. Neuroscience, in press

Mednick SA, Machon RA, Huttunen MO, Bonett D 1988 Adult schizophrenia following perinatal exposure to an influenza epidemic. Arch Gen Psychiatry 45:189–192

Nelson KB 1988 Perspective on the role of perinatal asphyxia in neurologic outcome. Can Med Assoc J (suppl) p 3–10

Reveley AM, Reveley MA, Clifford CA, Murray RM 1982 Cerebral ventricular size in twins discordant for schizophrenia. Lancet 1:540–541

Early stress and adult emotional reactivity in rhesus monkeys

Stephen J. Suomi

Laboratory of Comparative Ethology, National Institute of Child Health and Human Development, Building 31/Room B2B15, 9000 Rockville Pike, Bethesda, MD 20892, USA

Abstract. This chapter examines the relationship between early social experiences and behavioural and emotional reactivity in adolescence and adulthood that has been established through extensive research with rhesus monkeys. Classic studies carried out in the 1960s first demonstrated that rearing under conditions of social isolation resulted in severe behavioural abnormalities that carried over into adulthood. In the 1970s techniques for reversing such isolation-induced deficits were developed. More recent studies have examined the long-term consequences of more subtle variation in early rearing environments. Monkeys reared from birth without mothers but with extensive peer contact develop relatively normal social behavioural repertoires and function well in familiar and stable social settings. However, peer-reared monkeys display extreme behavioural and physiological reactions to environmental challenges, such as brief social separation, later in life. In contrast, monkeys reared by unusually nurturant foster mothers appear to develop effective strategies for coping with subsequent environmental challenges. Some general principles that have emerged from these studies with rhesus monkeys will be outlined and their implications regarding possible relationships between early social experiences and responses to challenge later in life in humans will be discussed.

1991 The childhood environment and adult disease. Wiley, Chichester (Ciba Foundation Symposium 156) p 171–188

The idea that certain aspects of the childhood environment can have important consequences for adult human functioning is central to many theories of social and personality development. For many years researchers and clinicians alike have searched for causal links between early experiences and adult behavioural and emotional characteristics. Such links have not been easy to establish empirically. While clinical reports suggesting powerful potential relationships between early experiences and later proclivities abound, the retrospective nature of most of these reports precludes the establishment of clear-cut causal links. Moreover, there is a dearth of relevant prospective data, due in no small part to the formidable ethical and practical obstacles that researchers inevitably face in designing and carrying out appropriately controlled human longitudinal studies that span the period from infancy to adulthood. As a result, theoretical speculations regarding possible long-term consequences of early experiences have largely outstripped the actual data.

On the other hand, there is a wealth of studies investigating the consequences of different early physical and social environments on biobehavioural development in many species of animals. Some of these studies have been carried out with non-human primate subjects such as rhesus monkeys (*Macaca mulatta*). This chapter examines the relationships that have been established between early social experiences and biobehavioural reactivity in adolescence and adulthood through extensive research with rhesus monkeys. The chapter begins with a review of classic studies investigating the consequences of early social isolation for subsequent social and emotional development, as well as experimental efforts to reverse the adverse consequences of such isolation rearing. More recent longitudinal research investigating the effects of more subtle differences in early rearing histories will then be examined. Some general principles that have emerged from these prospective studies with rhesus monkeys will be outlined, and their implications regarding possible relationships between early experiences and responses to environmental challenges later in life at the human level will be discussed.

Classic studies of early social isolation

Researchers who study biobehavioural development in non-human primates have long been aware that early experiences can have profound and prolonged physical, behavioural, and physiological consequences. Much of this knowledge can be traced to the extensive studies of the effects of early social isolation that were carried out in a variety of non-human primate species during the 1960s. For example, classic experiments by Harlow and his co-workers (Harlow et al 1965, Harlow & Harlow 1969), in which rhesus monkey infants were reared in a variety of social environments that differed systematically in the nature and extent of conspecific stimulation, revealed that subjects raised for at least their first six months in tactile isolation from all other monkeys were highly likely to exhibit a variety of long-term behavioural pathologies. These isolation-reared monkeys failed to develop species-normative patterns of affiliation, exploration, and play as infants and juveniles; instead, they displayed patterns of self-directed and stereotypic motor activity seldom exhibited by socially reared control subjects. Such abnormalities emerged even if the monkey infants had extensive visual, auditory, and olfactory contact with conspecifics. Moreover, as these isolation-reared monkeys matured they continued to display deficits in species-normative behaviour patterns, as well as unusually high levels of self-directed and stereotypic activity, even when given extensive subsequent experience with socially normal age-mates (Mitchell et al 1966, Mitchell & Clark 1968). As adolescents and adults they often exhibited excessive and/or inappropriate aggression toward other monkeys, they were generally incompetent sexually, and a high proportion of the females who became pregnant were inadequate or even abusive in their maternal behaviour, at least for their firstborn offspring (Harlow et al 1966).

It should be pointed out that most of the above behavioural abnormalities were *not* developed by rhesus monkeys who as infants were not completely isolated physically, but instead received limited (i.e., less than species-normative) social experience with a few conspecifics (e.g., peers). Moreover, there appeared to be substantial variability in the relative severity and permanence of isolation-induced abnormalities among different primate species, even those who share the same genus (Sackett et al 1976). Indeed, the early rhesus monkey studies revealed substantial *individual* differences in response to early isolation. For example, approximately one-third of all female isolation-reared rhesus monkeys displayed relatively normal maternal behaviour toward their firstborn and all subsequent offspring (Ruppenthal et al 1976).

At any rate, the early isolation work yielded a number of general principles, some of which can be summarized as follows. (1) The effects of early social isolation could be described in terms of *sensitive periods*—that is, isolation occurring early during the process of development had disproportionate effects in terms of the relative severity and duration of consequences compared to isolation occurring later in life. (2) The specific nature of early isolation-induced social deficits changed as a function of developmental stage. Thus, social deficits exhibited during adulthood were *qualitatively different* from those exhibited during infancy or adolescence. (3) The effects of early social isolation were not uniform either across or within species of non-human primate subjects. Instead, there were substantial *species and individual differences* in the relative severity of both short- and long-term consequences of early social isolation.

Social rehabilitation of isolation-reared monkeys

During the 1970s several studies demonstrated that most of the early isolation-induced behavioural abnormalities in rhesus monkeys could be reversed by exposing the isolates to socially competent infant partners. Suomi & Harlow (1972) reported dramatic improvements in the behaviour of rhesus monkeys socially isolated for their first six months of life and then exposed to socially normal three-month-old 'therapist' monkeys for daily one-hour periods of interaction. Before exposure to the younger 'therapists', the isolation-reared monkeys had displayed a wide array of self-directed and idiosyncratic stereotypic behaviour patterns, and in their initial interactions with the 'therapist' monkeys they failed to exhibit spontaneous play or any other socially directed behaviour. In contrast, the therapist monkeys readily initiated social contact with the isolates (just as they might initiate social contact with their own mothers), and the isolates soon began to reciprocate the contact. When this occurred, the isolate monkeys' incidence of self-directed and stereotypic behaviours dropped precipitously. Shortly thereafter, the therapist monkeys began to initiate play bouts with the isolates (as they normally would do with peers in species-normative settings), and once again the isolate monkeys reciprocated these social initiations. Within a few weeks the isolate monkeys began to initiate their own play bouts, and they subsequently displayed species-normative patterns of behavioural development as they grew up, with a relative lack of self-directed or stereotypic activity.

It is worth noting that the appearance of 'normal' patterns of social and non-social behaviour, and the great reduction of clearly abnormal patterns, generalized to other situations in which the 'therapist' monkeys were not present (cf. Suomi et al 1972). Moreover, these 'rehabilitated' isolation-reared monkeys continued to display normal behavioural repertoires as they passed through adolescence and into adulthood (cf. Cummins & Suomi 1976). On the other hand, when these monkeys encountered novel or stressful circumstances, their initial reactions typically included a brief reappearance of self-directed or stereotypic behaviour, although those previously displayed patterns usually subsided rather quickly. Thus, although the relative incidence of self-directed and stereotypic behaviour was greatly reduced in the rehabilitated isolates, some abnormal patterns remained in these monkeys' overall behavioural repertoires even into adulthood.

Novak & Harlow (1975) used similar techniques of exposure to three-month-old 'therapists' in a largely successful effort to rehabilitate rhesus monkey juveniles who had been socially isolated for their first 12 months of life. As was the case for the previously described six-month isolation-reared monkeys, these 12-month isolates displayed grossly abnormal behaviour repertoires prior to exposure to their younger, socially normal 'therapist' partners, and they similarly developed relatively normal behavioural repertoires through extensive interactions with those 'therapist' monkeys. Moreover, the 12-month isolates generalized their species-normative patterns of activity to other situations and toward other social partners as they passed through adolescence and into adulthood. On the other hand, many of their earlier patterns of aberrant self-directed and stereotypic behaviour would reappear briefly whenever these monkeys encountered novel or stressful circumstances (cf. Novak 1979). Thus, as was the case for the six-month isolates, these 12-month isolates retained some of the aberrant components of their behavioural repertoire into adulthood, even though such components were rarely exhibited and then usually limited to novel or stressful situations.

These and other experimental efforts to reverse the long-term consequences of early isolation rearing (e.g., Noble et al 1976) also yielded a number of general principles, some of which can be summarized as follows. (1) The most successful interventions were those designed specifically to reproduce the very social stimulus patterns (e.g., close physical contact) that the isolates had 'missed' during their period of isolation and to present these stimuli in a sequence simulating that of species-normative development, even if occurring substantially later chronologically than normal. (2) The most successful interventions were those initiated soon after the isolation-reared monkeys emerged from isolation. (3) The improvements in the behavioural repertoires of 'rehabilitated' isolates were most likely to be sustained if these monkeys were maintained in socially stimulating but stable environments. Return to socially barren environments typically resulted in a return of elevated levels of self-directed and idiosyncratic stereotypic behavioural patterns. Moreover, (4) circumstances that involved environmental novelty or challenge readily elicited the appearance of aberrant behavioural patterns previously developed during the period of early isolation, even if such patterns had not been exhibited in intervening periods spanning many months or even years. Thus, it seems that specific behaviour patterns, developed early in life, were never totally lost from a behavioural repertoire, even if not displayed for years. Rather, it appeared that there were certain environmental conditions capable of re-eliciting these specific behavioural patterns, be they normal or abnormal.

Long-term consequences of early peer rearing

The studies investigating long-term consequences of early social isolation carried out in the 1960s and early 1970s involved rearing conditions that differed dramatically from those experienced by any rhesus monkey infant growing up in the wild. Social isolation initiated at birth clearly represents a major departure from the species-normative pattern of early rearing by any standard. In contrast, recent primate studies of the consequences of different early experiences have utilized rearing environments that provide a much greater degree of social stimulation than did the early isolation experiments. For example, my colleagues and I have conducted a series of longitudinal investigations at the Laboratory of Comparative Ethology, National Institute of Child Health and Human Development, focusing on rhesus monkeys reared from birth away from their biological mothers but with extensive access to like-reared age-mates, comparing and contrasting their patterns of behavioural and physiological development with those of mother-reared infants.

The peer-reared rhesus monkeys were separated from their mothers at birth and hand-reared (with inanimate surrogate 'mothers') in the laboratory nursery for their first 30 days of life. At one month of age they were moved to a group cage containing 4–6 nursery-reared peers of both sexes, and they remained in the group cage with their peers until they were six months old. After a series of four short-term separations, during which the subjects were individually housed for four-day periods, the peer-reared monkeys were put into larger, mixed-sex groups containing 10–12 age-mates, some of whom had been peer-reared and some mother-reared. The mother-reared monkeys had spent their first six months living with their biological mothers in dyad cages, after which they also underwent a series of four, four-day-long separations before being moved into these larger social groups. Both the mother-reared and peer-reared monkeys continued to live in these groups through puberty and into early adulthood. The only experimental disruptions of these group living arrangements involved an annual series of four separations (i.e., at 18 months, 30 months, etc.) during which time each subject was individually housed for four-day-long periods, as had been the case for the original separations at six months. Thus, the peer-reared and mother-reared subjects differed only in terms of their respective rearing environments during their first six months of life. Thereafter, they lived in the same social groups and were subjected to the same short-term separations at the same chronological ages until they were young adults.

Consistent with reports from previous studies (e.g., Chamove et al 1973, Harlow 1969, Suomi 1979), the peer-reared monkeys exhibited relatively normal behavioural development within their rearing groups. Unlike isolation-reared infants, they failed to show chronically high levels of either self-directed behaviour (except for thumb-sucking) or idiosyncratic stereotypies, but instead developed a wealth of species-normative, socially directed activities, including exploration, social grooming, play, and age-appropriate sex posturing. Peer-reared monkeys did differ somewhat from their mother-reared counterparts early in life, in that they reduced their levels of infantile partner-directed clinging more slowly over time than mother-reared infants reduced their levels of mother-directed clinging, and they were somewhat slower to develop complex play patterns than were their mother-reared counterparts. In addition, as infants and juveniles, peer-reared monkeys appeared to be more timid than mother-reared youngsters, in that they were more reluctant to explore new surroundings or initiate interactions with unfamiliar age-mates (Higley & Suomi 1989). However, such differences could be described as basically quantitative rather than qualitative, unlike the case for isolation-reared monkeys. Moreover, most of the differences in levels of specific behaviour patterns between peer-reared and mother-reared subjects essentially disappeared as these monkeys matured. As adolescents and young adults, the peer-reared subjects displayed age-appropriate levels of play, aggression, and sex within their familiar social groups, and individual peer-reared monkeys were as successful in attaining high-ranking positions in the group dominance hierarchy as were their mother-reared counterparts.

In contrast, dramatic differences emerged between mother-reared and peer-reared subjects in both behavioural and physiological responses to the social separation manipulations each year. Numerous studies over the past quarter-century have demonstrated that social separation is a potent stressor for rhesus monkeys of all ages, typically resulting in substantial behavioural disruption and physiological arousal for at least the initial hours of separation (cf. Mineka & Suomi 1978 or Suomi 1982 for a comprehensive review of this topic). However, during each series of annual separations the peer-reared monkeys exhibited far more extreme reactions than did their mother-reared counterparts, even when they were being separated from the same social groups. During the separations at six months of age, peer-reared monkeys displayed significantly higher levels of self-directed behaviours, had more distress vocalizations, and were more passive than were mother-reared subjects. Peer-reared monkeys also responded to separation with greater activation of the hypothalamic-pituitary-adrenal axis, as indicated by higher levels of plasma cortisol and adrenocorticotropic hormone (ACTH), and they showed greater turnover of the noradrenergic system during separation, as evidenced by lower cerebrospinal fluid levels of noradrenaline and higher levels of 3-methoxy-4-hydroxyphenylglycol (MHPG), a noradrenaline metabolite.

During the separations at 18 months of age, the above-described rearing condition differences in behavioural and physiological reactions were as great as or greater than they were during the six-month separations. In addition, during the 18-month separations peer-reared subjects displayed significantly higher cerebrospinal fluid levels of 5-hydroxyindoleacetic acid (5-HIAA), a serotonin metabolite, than mother-reared subjects (there were no rearing condition differences in 5-HIAA levels during the six-month separation series). Moreover, the same basic pattern of more extreme behavioural and physiological response to separation in peer-reared monkeys was continued during subsequent annual separations (Higley et al 1990). On the other hand, there were few, if any, significant rearing condition differences during periods of stable group housing as the monkeys passed through puberty and into early adulthood.

Thus, while mother-reared and peer-reared monkeys showed increasingly similar behavioural and physiological profiles within their common social groups as they grew older, their respective responses to the stressful challenge of repeated separation remained different, or became even more so, over the same period of development. What might account for these distinctive patterns of response to separation? One plausible explanation comes from consideration of Bowlby's attachment theory (Bowlby 1969, 1973). It can be argued that one fundamental way in which the mother-reared and peer-reared monkeys differed during infancy was in the nature of the *attachment relationships* that they formed with their respective rearing partners. Peer-reared infants clearly establish attachment-like bonds with specific partners that appear to mimic species-normative mother–infant bonds (cf. Suomi 1979). However, data from a variety of sources suggest that an infant peer of either sex is not nearly as effective an attachment object as is an adult female in two fundamental respects: (a) ability to serve as a secure base for exploring novel objects, partners, or situations; and (b) ability to comfort an infant and reduce its fear in the face of challenge (e.g., Meyer et al 1975, Higley et al 1988). Thus, in attachment theory terms, peer-reared monkeys are much more likely to form *insecure* attachment relationships during their first six months of life than are mother-reared individuals. While such differences in the nature of early attachments would be expected to be of little consequence when individuals are in stable, familiar physical and social settings, according to the theory, insecure early attachments should predispose individuals to display extreme disruption under conditions of stress or challenge, especially those involving some form of social loss (cf. Bowlby 1973, Sroufe & Waters 1977).

This, of course, is precisely the pattern of results observed in the above-described studies comparing peer-reared and mother-reared rhesus monkey biobehavioural development: there were relatively minor behavioural and physiological differences in developmental patterns during the six-month period of differential rearing, and those differences diminished thereafter during subsequent periods of stable group living. In contrast, there were dramatically different initial patterns of biobehavioural response to the challenge of social separation—differences that if anything became exaggerated as the monkeys grew older, long after the differential early rearing experiences. In other words, under stable, non-stressful conditions there was developmental discontinuity from early rearing histories to later behavioural and physiological functioning, but under periodic conditions of challenge there was striking developmental continuity in both behavioural and physiological response.

Individual differences and foster-mother rearing

The strong developmental continuities in response to environmental challenge seen in the above-described studies of mother- vs. peer-reared monkeys have provided the basis for another set of studies investigating the relationship between different early social environments and later functioning. Recent work in my laboratory has demonstrated major developmentally stable *individual differences* among like-reared monkeys in their responses to environmental challenges. For example, approximately 20% of the mother-reared rhesus monkeys in our laboratory colony react to brief social separations with unusually high cortisol and ACTH elevations, exaggerated noradrenaline turnover, and much more 'depressive' behavioural reactions than the other mother-reared subjects. These 'high-reactive' monkeys, as we call them, also display extreme behavioural and physiological reactions in other novel and/or challenging settings. For example, when first placed in a playroom full of toys and climbing apparatus, most four-month-old monkeys will soon begin to explore the playroom and manipulate the toys. Four-month-old high-reactive monkeys, in contrast, typically stay in a corner of the playroom over the same time period, and during that period they are also more likely to have higher levels of cortisol and ACTH, and higher and more stable heart rates, than their less reactive counterparts. Thus, circumstances that elicit exploration and manipulation in most monkeys instead elicit fearful, 'anxious' responses and heightened physiological arousal in high-reactive subjects (Suomi 1987, 1991).

My colleagues and I have been interested in these dramatic individual differences for several reasons. First, not only are these different patterns of biobehavioural response remarkably consistent across different types of challenge, but they are also exceedingly stable developmentally, and they are apparent very early in life, certainly within the first month. Second, we have found parallel patterns of individual differences in biobehavioural response to challenge in troops of monkeys living in the wild—in terms of relative proportions of high-reactive individuals, of characteristic physiological profiles, and of the stability of response patterns over time (Rasmussen & Suomi 1989, Suomi et al 1989). Third, we have been struck by the degree to which our high-reactive young monkeys resemble human children identified as shy or 'behaviourally inhibited', in terms of their characteristic behavioural and physiological response to environmental novelty and challenge, as well as the developmental stability of the respective phenomena (cf. Kagan et al 1988).

An increasing body of data, albeit largely circumstantial in nature, strongly suggests that these differences in response to challenge are highly heritable. In my laboratory we have identified members of the breeding colony who have consistently produced high-reactive offspring—and other breeding pairs who have never had high-reactive infants (cf. Suomi 1987). On the other hand, the previously described studies comparing mother- and peer-reared monkeys indicated that response to challenge can also be influenced by early rearing experiences, especially those involving different attachment relationships. We therefore sought to determine whether high-reactive monkeys might be differentially affected by different attachment relationships, relative to less reactive individuals.

In order to do this, we carried out a study in which rhesus monkey infants, selectively bred to be high reactive, were foster-reared by multiparous females who differed in their characteristic maternal 'style', as determined by their care of previous offspring. Some of these foster mothers had consistently demonstrated unusually 'nurturant' maternal care, in that they were especially protective and supportive of their infants' exploratory efforts, and much less punitive and rejecting during weaning, than the norm for the species. Thus, infants known to be predisposed (on the basis of pedigree) to become high reactive were reared either by unusually nurturant foster mothers (beginning in the first week of life) or by control foster mothers. Other infants, specifically bred not to be predisposed to high reactivity, were likewise cross-fostered within the first week of life either to unusually nurturant females or to 'normal' foster mothers. We then watched these different infants grow up with their respective foster mothers (Suomi 1987).

During the first six months of life the high-reactive infants cross-fostered to control foster mothers generally showed normal patterns of behavioural development. There were relatively few differences in behavioural levels between these infants and the low-reactive infants cross-fostered to either control or nurturant foster mothers (which likewise did not differ from each other). Somewhat surprisingly, however, high-reactive infants cross-fostered to nurturant females actually displayed somewhat accelerated development—they showed reduced levels of physical contact with their foster mothers at earlier ages, had higher levels of locomotion and exploration away from their foster mothers, and displayed less behavioural disturbance during weaning than did high-reactive infants cross-fostered with control mothers or low-reactive infants reared by either type of foster mother. Thus, the high-reactive infants seemed to benefit more from rearing by a nurturant foster mother than did low-reactive infants.

However, when the high- and low-reactive infants were briefly separated from their foster mothers at six months of age (using the same separation paradigm as in the previously described studies of early mother vs. peer rearing), the high-reactive infants displayed more severe behavioural and physiological reactions than their low-reactive counterparts, regardless of the type of foster mother they had been reared with. On the other hand, as soon as the separations were over, high reactive individuals with nurturant foster mothers once again appeared behaviourally precocious, relative to high-reactive infants with control foster mothers and low-reactive infants with either type of foster mother.

A largely unexpected and important pattern emerged when these monkeys were moved into larger peer groups in succeeding months, as had been done in the previous studies of mother vs. peer rearing. These larger peer groups differed from those previously described, however, in that they also contained an old male–female pair (colloquially called 'foster grandparents'). The foster grandparents were added in order to facilitate the socialization of peer group members—the old male effectively broke up fights and controlled the more rambunctious individuals in the group, while the old female was available for those youngsters in the group desiring close physical contact.

In this larger group setting, high-reactive monkeys who had been reared by a nurturant foster mother immediately established close social relationships with these old adults, particularly the old female. With the old adults providing a basis of social support, these high-reactive youngsters became the most dominant members of the peer group. In contrast, high-reactive monkeys who had been reared by control foster mothers avoided the older monkeys and wound up at the bottom of the dominance hierarchy. Low-reactive individuals, independent of the kind of foster mothers with whom they had been reared, also did not establish close relationships with the older animals, and they achieved intermediate positions within the dominance hierarchy of their social group. Perhaps of greatest interest was the fact that the high-reactive monkeys reared by nurturant foster mothers retained their elevated dominance status thereafter, even after the foster grandparent pair was removed from their social group a year later. These findings suggest that specific types of early social experiences (e.g., being reared by an unusually nurturant foster mother) may have different long-term consequences for individuals with different predispositions.

Conclusions and implications

The results of these more recent studies, involving less extreme manipulations of early rearing environments than in the classic social isolation studies, have yielded a number of principles that are largely congruent with those from the classic studies, as well as providing some additional insights. First, it seems clear that different sets of early social experiences can be linked to specific behavioural and physiological propensities expressed later in life. Moreover, it appears that the more such early experiences deviate from species-normative patterns, the more extreme will be the long-term consequences. In this regard, early experiences concerning attachment relationships (or lack thereof) seem especially relevant for subsequent social-emotional development.

Second, the long-term consequences of different sets of early social experiences are most likely to be expressed under conditions involving some degree of novelty or challenge—situations that are stressful and tend to elicit emotional reactions. In the absence of stress, behavioural and physiological patterns attributable to differential early experiences are likely to be masked. Indeed, in stable, relatively benign environments, adverse early experiences may appear to have no readily discernible long-term consequences.

Finally, it appears that the long-term consequences of specific early social experiences may not be uniform across all individuals. Rather, some individuals, perhaps due to their heritable predispositions, may be more sensitive to the effects of particular early experiences than are others. Thus, not all female monkeys isolated in those early experiments exhibited subsequent deficits in their maternal behaviour. More recent data indicate that high-reactive infants are more likely to be influenced by the maternal style of their caretaker than are low-reactive infants, with different long-term consequences as well.

What implications do these findings from studies of rhesus monkeys have for the possible effects of early experiences on subsequent functioning in humans? Of course, rhesus monkeys are not furry little humans with tails, but instead members of another primate species. Moreover, few if any humans are likely to have experienced the extreme degree of early social deprivation that characterized some of the early monkey isolation studies. Thus, it is unlikely that all of the specific findings from these monkey studies will generalize point-by-point to the human case.

On the other hand, it seems likely that most, if not all, of the more general principles that can be gleaned from these monkey data can, in fact, apply to the human case. A wealth of data suggests that early attachment relationships do have long-term consequences in humans, especially under conditions of stress and challenge (cf. Bowlby 1973, Sroufe & Waters 1977). Moreover, the pattern of individual differences in the behavioural and physiological response to challenge observed in rhesus monkeys seems strikingly similar to the pattern of individual differences reported for human infants and young children (cf. Kagan et al 1988). Thus, in considering possible long-term consequences of early experiences in humans, one might do well to keep in mind the lessons from the monkey literature. These lessons are that early experiences that deviate greatly from the norm are those most likely to have long-term impact, that the most dramatic long-term effects will be most likely to be seen under conditions of stress, and that there may well be substantial individual differences in how specific early experiences affect later behavioural and emotional proclivities.

References

Bowlby J 1969 Attachment. Basic Books, New York
Bowlby J 1973 Separation: anxiety and anger. Basic Books, New York
Chamove AS, Rosenblum LA, Harlow HF 1973 Monkeys (*Macaca mulatta*) raised only with peers: a pilot study. Anim Behav 21:316–325
Cummins MS, Suomi SJ 1976 Behavioural stability of rhesus monkeys following various rearing. Primates 17:42–51.
Harlow HF 1969 Age-mate or peer affectional system. In: Lehrman DS, Hinde RA, Shaw E (eds) Advances in the study of behavior. Academic Press, New York, p 21–38
Harlow HF, Harlow MK 1969 Effects of various mother-infant relationships on rhesus monkey behaviors. In: Foss BM (ed) Determinants of infant behaviour, vol 4. Methuen, London, p 15–36
Harlow HF, Dodsworth RO, Harlow MK 1965 Total social isolation in monkeys. Proc Natl Acad Sci USA 54:90–96
Harlow HF, Harlow MK, Dodsworth RO, Arling GL 1966 Maternal behaviour of rhesus monkeys deprived of mothering and peer associations in infancy. Proc Am Philos Soc 110:58–66
Higley JD, Suomi SJ 1989 Temperament in nonhuman primates: reactivity, its description, and correlates with psychopathology. In Kohnstamm GA et al (eds) Handbook of temperament in children. Wiley, New York, p 153–167
Higley JD, Danner GR, Hirsch RM 1988 Attachment in rhesus monkeys reared with only peers or with their mothers as assessed by the Ainsworth Strange Situation procedure. Infant Behav Dev 11:139–141
Higley JD, Suomi SJ, Linnoila M 1990 Developmental influences on the serotonergic system and timidity in the nonhuman primate. In: Coccaro EF, Murphy DL (eds) Serotonin in major psychiatric disorders. American Psychiatric Press, Washington, DC, p 29–46
Kagan J, Renick JS, Snidman N 1988 Biological basis of childhood shyness. Science (Wash DC) 240:167–171
Meyer J, Novak MA, Bowman RE, Harlow HF 1975 Behavioral and hormonal effects of attachment object separation in surrogate-peer-reared and mother-reared infant rhesus monkeys. Dev Psychobiol 8:425–435
Mineka S, Suomi SJ 1978 Social separation in monkeys. Psychol Bull 85:1376–1400
Mitchell GD, Clark DL 1968 Long term effects of social isolation in nonsocially adapted rhesus monkeys. J Genet Psychol 113:117–128
Mitchell GD, Raymond EJ, Ruppenthal GC, Harlow HF 1966 Long term effects of total social isolation upon behavior of rhesus monkeys. Psychol Rep 18:567–580
Noble AB, McKinney WT, Mohr C, Moran E 1976 Diazepam treatment of socially isolated monkeys. Am J Psychiatry 133:1165–1170

Novak MA 1979 Social recovery of monkeys isolated for the first year of life. II. Long-term assessment. Dev Psychol 15:50–61

Novak MA, Harlow HF 1975 Social recovery of monkeys isolated for the first year of life. I. Rehabilitation and therapy. Dev Psychol 11:353–365

Rasmussen KLR, Suomi SJ 1989 Heart rate and endocrine responses to stress in adolescent male rhesus monkeys on Cayo Santiago. Puerto Rican Health Sci J 8:65–71

Ruppenthal GC, Arling GL, Harlow HF, Sackett GP, Suomi SJ 1976 A 10-year perspective on motherless mother monkey behavior. J Abnorm Psychol 85:341–349

Sackett GP, Holm R, Ruppenthal GC 1976 Social isolation rearing: species differences in the behavior of macaque monkeys. Dev Psychol 10:283–288

Sroufe LA, Waters E 1977 Attachment as an organizational construct. Child Dev 48:1184–1192

Suomi SJ 1979 Peers, play, and primary prevention in primates. In: Kent M, Rolf J (eds) Primary prevention of psychopathology, vol 3. New England Universities Press, Hanover, p 127–149

Suomi SJ 1982 Abnormal behavior and animal models of psychopathology. In: Fobes JL, King JE (eds) Primate behavior. Academic Press, New York, p 171–215

Suomi SJ 1987 Genetic and environmental contributions to individual differences in rhesus monkey biobehavioral development. In: Krasnegor N, Blass E, Hofer M, Smotherman W (eds) Perinatal development: a psychobiological perspective. Academic Press, New York, p 397–420

Suomi SJ 1991 Primate separation models of affective disorders. In: Madden J (ed) Adaptation, learning and affect. Raven Press, New York, in press

Suomi SJ, Harlow HF 1972 Social rehabilitation of isolate-reared monkeys. Dev Psychol 6:487–496

Suomi SJ, Harlow HF, McKinney WT 1972 Monkey psychiatrists. Am J Psychiatry 128:927–932

Suomi SJ, Scanlan JM, Rasmussen KLR et al 1989 Pituitary-adrenal response to capture in Cayo-derived M-troop rhesus monkeys. Puerto Rican Health Sci J 8:171–176

DISCUSSION

Lloyd: Your picture of an isolated infant monkey in a cage reminded me very poignantly of the child who is abused or emotionally neglected. We know that such children are likely, when they grow up, to display behaviour to their own children similar to that they received from their mothers. You said that monkeys reared in isolation, even when they had been rehabilitated, behave abnormally under stress. How do they behave as mothers?

Suomi: Quite a lot is known about this topic. Of those females who were reared under conditions of deprivation of social contact from conspecifics for at least the first six months of life, whether they had surrogate mothers, whether they could see or hear other animals, or whether they were isolated visually as well as physically, the large majority (about 70%) showed either abusive behaviour or what Harlow called indifferent behaviour towards their firstborn offspring. They failed to retrieve these infants or to nurse them, and thus intervention by staff was required to sustain the infants. The other 30% of those original isolated females spontaneously showed species-normative maternal behaviour without any intervention or rehabilitation, so there were individual differences.

Very interestingly, a large proportion of the previously isolated females who were incompetent as mothers towards their firstborn offspring turned out to be very good mothers to their subsequent offspring. We carried out a retrospective study looking at conditions that would facilitate this (Ruppenthal et al 1976). One crucial factor was how much time the previously isolated mothers actually spent with their first infant. If these females had spent at least 48 hours with their firstborn infant before it had to be removed for its own survival, then their likelihood of being competent in their care toward subsequent offspring was nearly 100%. If they spent fewer than 48 hours with that first infant and it was then removed, the likelihood that they would be neglectful of or would abuse a subsequent offspring also approached nearly 100%. This suggested that there are substantial individual differences in the degree to which maternal behaviour is affected by early social isolation, particularly with respect to contact-oriented activity. Nevertheless, a small amount of contact with an infant might serve as a primer, to get maternal behaviour going, and such contact would lead to relatively normal maternal behaviour toward subsequent infants. So substantial recovery of maternal behaviour was possible in the early isolation-reared females without formal early rehabilitation.

Lloyd: But did improvement during infancy and after the period of isolation correlate with whether they were good mothers when they had their *first* infant?

Suomi: Virtually all those females who had been isolated and then rehabilitated via exposure to younger therapists were good mothers, even with their first baby. In fact, some of those females are in our breeding colony now.

I should add that when females who were normally reared, either by their own mothers or in peer groups, but who subsequently showed the most severe reactions to brief separation (i.e., who were in our high-reactive group), were compared with females who had the same separation experiences but whose reactions to the separations were mild (i.e., who were in our low-reactive group), there were major differences in their behaviour toward their firstborn offspring. The females in the low-reactive group were all perfectly good mothers. In contrast, almost 80% of the females who had early separations and showed depressive reactions to them failed to take care of their firstborn offspring. The problems for these high-reactive females occurred in the first two weeks of their infant's life; after that early period they displayed normal maternal behaviour. So again, there seems to be something rather important about the early post partum period.

The veterinary practice at that time was to house females individually, pairing them with a male during their periods of ovulation and then maintaining them in individual cages during their pregnancy and period of delivery. This is very different from what happens in the wild, where each monkey female becomes pregnant within her social group and delivers her infant surrounded by her female relatives. When we saw these results we changed our procedures and started maintaining our high-reactive females in social groups during pregnancy. When we did that, the problem of inadequate maternal behaviour disappeared.

Hamosh: Do you have any data on the prolactin levels of the good and bad mothers? Prolactin is a hormone that facilitates mothering.

Suomi: We do not. Such data would not be difficult to obtain, and we probably have blood samples from some of these high-reactive females. It would be a relatively simple thing to do.

Hamosh: Do you also have data on milk volume and the let-down reflex? Is there also something wrong with the initiation of lactation in the less competent mothers? This could be mediated by prolactin as well.

Suomi: No, we don't have these data. Incidentally, it is not a comparison of 'good' and 'bad' mothers. For example, our cross-fostering data basically compare mothers who show unusual support of exploration by their young and unusually low levels of rejection and punishment of their offspring at the time of weaning, relative to the normal control mothers. Thus, the comparisons in the cross-fostering study are really of unusually nurturant mothers versus mothers showing species-normative levels of these various behaviours.

Lucas: We have been talking, in this meeting, about adverse outcomes, which are to do with medical failures, as it were, in life. Social success is also a very important outcome. Two possible manifestations of 'success' in humans are power and money. Interestingly enough, many of the people who have markedly achieved those, in the UK at least, have experienced a rather stereotyped series of events: they have been rejected by both their parents, and sent away to boarding school, where they have been brought up with peers who have also been rejected by their parents; they have been looked after by surrogate mothers; and they have then emerged as leaders! For instance, over a third of the 200 richest people in Britain come from just one boarding school. This isn't an entirely frivolous remark, because your data on those who took up leadership of the monkey group actually conformed not too dissimilarly to that pattern.

Suomi: I'll try to give a non-frivolous answer! My purpose in presenting those data was to suggest that even when one can identify, as I believe we have, highly heritable predispositions, they clearly are subject to environmental modification. Thus, a better way to predict outcome is to know (a) the presumed pedigree of the individuals, (b) the early experiential circumstances under which they were reared, and (c) the current situation, that allows them either to take advantage of those early rearing experiences or to be at an even greater risk of extreme reactions to novelty and challenge because of events that may have happened early in childhood.

It is not difficult to hypothesize various pathways, both behavioural and physiological, by which monkeys who show unusual reactions to novelty or challenge can learn not only to cope with such situations, but also to take advantage of their extra interest and extra physiological arousal. For example, individual rhesus monkeys whose response to novelty is to show sustained sympathetic nervous system arousal and adrenocortical activity may be able to be more vigilant defenders of their offspring; they may be able to sustain an aggressive interaction for a longer period of time; or they may be able to direct more attention towards a novel object or a potentially dangerous situation than other individuals who may be less reactive to the same circumstances. The one thing that characterizes our highly reactive animals is that they are more responsive to both positive and negative changes in their environment than their counterparts.

Rutter: I have two questions on your fascinating set of studies on the interplay between genetic and environmental factors. Both concern the mechanisms involved in the perpetuation of the effects of experiences.

The first refers back to the early Harlow studies (Ruppenthal et al 1976). They began at a time when people were interested in the permanent effects of early experiences and the results seemed to show that the effects were extremely lasting. Your subsequent work (Suomi et al 1974) and that of Novak (1979) modified that conclusion considerably. To what extent do you think the perpetuation of effects of early isolation was a consequence of the monkeys acting in ways that prevented them from having remedial experiences, if I can put it that way?

Suomi: I think the results had an enormous amount to do with that. One reason we chose, in our initial 'therapy' studies, to select social partners who were younger and socially naive to interact with the disturbed individuals was because if the isolation-reared monkeys were paired with age-mates, these age-mates would attack the isolates, because a normal rhesus monkey's reaction to a stranger is to display aggression against it, and the isolates certainly appeared strange. That kind of social stimulation would hardly be conducive to recovery! So we tried to choose, as therapists, individuals who were too young to display aggression, and who instead would show contact-directed activity like that directed toward a mother, and whose initial social behaviours would be simple enough not to overwhelm the isolates, who were behaviourally naive. When provided with that kind of stimulation, the isolates did very well.

Of similar interest, despite showing the full range of aberrant behaviour typical of isolates prior to therapy, including idiosyncratic stereotyped behaviour, when these animals were reintroduced to an isolation condition *after* the recovery via 'therapy', those aberrant behaviour patterns would come back. They were still in the rehabilitated isolate's repertory but were no longer elicited in familiar social circumstances. This suggests that one must pay attention to the circumstances in which these individuals are living, and in which they have been exposed, subsequent to an intervention.

Rutter: These findings illustrate the artificiality of making an absolute distinction between permanent and transient effects (Rutter 1989).

My second question concerns the possibility of differentiating between genetically and experimentally determined behavioural inhibition. Are there differences, either physiological or behavioural, in the form of behavioural inhibition according to its origin?

Suomi: We are trying to find that out now. Let me rephrase your question. Is it possible to find circumstances in which a genetically not at-risk individual will show some of these high-reactive, stress-elicited behaviours? We have some evidence for this in some individual monkeys. Thus peer rearing pushes the threshold for this pattern of response to challenge substantially in one direction, so that individuals who might not have been classified as highly reactive, had they been mother reared, now showed patterns that we classified as high-reactive.

Murray: There is an enormous industry among behaviour geneticists of giving out personality questionnaires to twins and their relatives! These studies nearly always seem to suggest that personality in humans is determined by genes and specific environment; they rarely show a significant common environmental effect on personality, namely the sort of effect which you are showing so emphatically in your monkeys. How can we bring together your findings with these human findings?

Suomi: We see a substantial range of variance in the area of infant temperament within our rhesus monkey population. I would suggest that we happen to have found one temperamental dimension (namely, response to novelty or challenge) that seems to be different from and more stable than the rest of the temperament-like dimensions in our monkey groups. If we were studying aggressive behaviour or sociability we might not find the same degree of heritability or developmental stability that we are picking up in terms of response to novelty or challenge. So I don't think that these different aspects of personality are necessarily identical in terms of their respective heritability, on the one hand, and susceptibility to modification by experience, on the other. There may well be essentially a different 'set' for each temperament dimension. I would suggest that these various factors have been selected for differentially, and that different combinations of genes and experiences lead to these various patterns of individual differences. We just happened to pick one particular dimension which shows up consistently throughout development but which also can be modified through certain experiences.

Rutter: It seems to me that the behaviour geneticists are looking at personality variations within the normal range and with environmental conditions that exclude extremes. Steve Suomi's results have a close parallel in the findings regarding genetic and environmental effects on human intelligence. Cross-fostering designs, with samples focusing on contrasts between children reared in privileged and in disadvantaged environments, show quite marked effects from broad environmental variables such as the social class of the parents (Capron & Duyme 1989, Schiff & Lewontin 1986), yet this is not evident from genetic studies focusing on variations within the normal range (Plomin & Daniels

1987). Thus, part of the explanation seems to be that the effects of extreme and more ordinary environments may not be the same.

A second part concerns the measurement of behavioural differences. Steve Suomi's findings apply to behavioural features that are made manifest only in particular challenge situations. Personality questionnaires, in contrast, do not specify situations at all, but rather focus on usual or habitual patterns of behaviour.

Mott: Dr Suomi, could you say more about how you analyse the huge effects of sire on cortisol levels in the progeny, and about how big the interaction of the sire group is with the type of experience during rearing?

Suomi: All of the monkeys in the laboratory studies that I described are 'home grown'; essentially we have a colony of monkeys born in captivity for which we know the identity of both parents. Some of the subjects came from breeding groups where there was no deliberate selection of parental pedigree, and some came from carefully selected paired matings. As we started seeing sire effects on adrenocortical and other responses to challenge, we began to select parents deliberately, both sires and dams, who had a history of producing a greater than chance probability of certain kinds of offspring. When we now re-pair those individuals we get about an 80% 'success' rate; that is, if we try to produce a high- or low-reactive infant via selection of biological parents, we actually get the desired type of infant about 80% of the time. In the studies comparing mother and peer rearing, we have tried to have equivalent numbers of high- and low-reactive infants in each rearing group. We find that when the pedigree of the biological parents with respect to reactivity is balanced across rearing groups, a large experiential effect—higher cortisol levels following challenge in peer-reared than in mother-reared infants—is obtained.

References

Capron C, Duyme M 1989 Assessment of effects of socio-economic status on IQ in a full cross-fostering study. Nature (Lond) 340:552–554

Novak MA 1979 Social recovery of monkeys isolated for the first year of life. II. Long-term assessment. Dev Psychol 15:50–61

Plomin R, Daniels D 1987 Why are children in the same family so different from one another? Behav Brain Sci 10:1–15

Ruppenthal GC, Arling GL, Harlow HF, Sackett GP, Suomi SJ 1976 A 10-year perspective on motherless mother monkey behavior. J Abnorm Psychol 85:341–349

Rutter M 1989 Pathways from childhood to adult life. J Child Psychol Psychiatry Allied Discip 30:23–51

Schiff M, Lewontin R 1986 Education and class: the irrelevance of IQ genetic studies. Clarendon Press, Oxford

Suomi SJ, Harlow HF, Novak MA 1974 Reversal of social deficits produced by isolation rearing in monkeys. Hum Evol 3:527–534

Childhood experiences and adult psychosocial functioning

Michael Rutter

MRC Child Psychiatry Unit, Institute of Psychiatry, De Crespigny Park, Denmark Hill, London SE5 8AF, UK

Abstract. Various studies have shown statistically significant associations between adverse experiences in childhood and abnormal psychosocial functioning in early adult life. It should not be assumed that this finding necessarily means an enduring effect of early experience. Part of the explanation is that adverse environments tend to be persistent and, hence, that what is being reflected is simply continuity in risk factors. But even when such continuity is taken into account, substantial associations over time remain. Using data from two long-term longitudinal studies a variety of possible mediating mechanisms are considered and shown to be operative. These include: an immediate effect leading to emotional/behavioural disturbance in childhood that then persists into adult life (this mechanism may be more important than appreciated hitherto, because heterotypic continuity has concealed the strength of the persistence of disturbance); one risk environment increasing the likelihood of occurrence of a second, different risk environment; the establishment of patterns of behaviour that bring about later risk environments; and the development of an increased vulnerability to later risk environments.

1991 The childhood environment and adult disease. Wiley, Chichester (Ciba Foundation Symposium 156) p 189–208

During the last four decades there have been several marked swings of the pendulum in prevailing views on the extent to which adverse experiences in childhood have a lasting effect on psychosocial functioning in adult life (Rutter 1989). During the 1950s there were claims that deprivation during the preschool years usually led to irreversible ill-effects. Empirical research findings during the 1960s and 1970s negated those views and there were counter-arguments that long-term sequelae were rare in the absence of later stresses and adversities (Clarke & Clarke 1976). More recent evidence has demonstrated a complex mix of continuities and discontinuities (Rutter 1987) and investigators have come to focus on possible reasons for the diverse straight and devious pathways leading from childhood experiences to adult behaviour (Robins & Rutter 1990, Rutter 1989). It has come to be appreciated that there are several quite different ways, involving contrasting mechanisms, by which early experiences can have enduring psychosocial sequelae.

Initially it had seemed obvious that it was important to exclude the possible impact of later environments in any investigation of the long-term effects of early adversities (Rutter 1981). However, it came to be realized that this constituted a potentially misleading research strategy because it ruled out two important possible mechanisms. First, early experiences might operate by causing people to behave in ways that made it more likely that they would encounter later stresses and adversities. People's life experiences are not random occurrences; to a considerable extent we select and shape our environments (Scarr & McCartney 1983). For example, Robins' (1966) classic follow-up study of children who attended child psychiatric clinics showed that antisocial youngsters were much more likely than controls to experience disrupted relationships, marital break-down, unemployment and frequent changes of jobs—events generally regarded as important adult stressors. It could be that negative experiences in early life operate in part by inducing disruptive or other behaviours that markedly increase the *incidence* of later stressors and adversities which, in turn, predispose to psychopathology in adult life.

Second, early influences might operate by increasing individual vulnerability to later adversities. It is clear that there are marked individual differences in people's responses to environmental hazards (Rutter & Pickles 1991) and it may be that part of this variation in susceptibility is a consequence of prior experiences. For example, Rodger (1990) found that a childhood vulnerability index (derived from a combination of childhood adversities and emotional disturbance) was associated with a marked increase in the likelihood of developing an acute depressive disorder following adult stressors.

The first possibility emphasizes the need to consider adult environments as possible outcome measures, or dependent variables, following childhood experiences. The second requires examination of the possibility that long-term sequelae reside in an increased vulnerability to psychosocial hazards and, hence, that adult psychosocial functioning will be affected only if and when such hazards arise. In so far as either mechanism is operative, any overall estimate of the proportion of population variance explained by early experiences will severely underestimate their importance in subsamples of the population (Pickles & Rutter 1991, Rutter & Pickles 1991).

These two possibilities, of course, represent only a small selection of possible pathways from childhood experiences to adult psychosocial functioning (see Rutter 1989 for others). Nevertheless, they do bring out three crucial considerations that constitute the main focus of this paper. First, many long-term sequelae depend for their occurrence on several, sometimes many, intervening links in a chain of indirect connections. When such links are all present, the cumulative long-term effects may be quite strong; but, in their absence, there may be no enduring consequences of even severe early adversities. Second, these links reflect a varied range of mechanisms each of which must be studied in its own right. This need is particularly great because there are important sex differences in the role of some crucial links. Third, because long-term effects so often depend on indirect chain effects reflecting varied mechanisms, there are many turning points at which life paths may exhibit a change of trajectory. These features will be discussed in relation to the findings from two rather different longitudinal studies spanning the period from early or middle childhood to adult life.

Two longitudinal studies

First, there is Quinton's long-term follow-up of 171 children who, because of a breakdown in parenting, were admitted in early life to Children's Homes run on group cottage lines (Pickles & Rutter 1991, Quinton & Rutter 1988, Rutter et al 1990, Zoccolillo et al 1991). In this case the risk factor in childhood comprised the family discord and disruptive parenting that preceded the admission to the Group Home, plus the experience of an institutional upbringing. This sample was compared with a quasi-random general population sample of 83 children living with their own families in the same broad area of inner London. Both groups were assessed for childhood conduct disturbance by means of contemporaneously completed standardized teachers' questionnaires and Juvenile Court records. Follow-up data in early adult life derived from standardized investigator-based interviews with subjects and their spouses lasting some two to four hours. Psychosocial functioning was systematically assessed with respect to the person's performance in key social domains such as work, friendships and love relationships. For the sake of brevity, this investigation will be referred to as the 'in care' study (because admission to the Group Homes involved being taken into the care of the Local Authority).

The second investigation to be discussed concerns the National Child Development Study (NCDS) which Ghodsian used to investigate the childhood antecedents of emotional disturbance at 23 years of age (M. Ghodsian, unpublished data 1989; M. Rutter, invited paper, Tenth Biennial Meeting of the International Society for the Study of Behavioural Development, Jyvaskyla, Finland 1989). It comprises a longitudinal investigation of some 17 000 children born during one week in 1958. For present purposes, socio-economic disadvantage in childhood (a composite measure based on a variety of indices of family circumstances such as financial hardship, paternal unemployment, and a lack of household amenities) is used as the childhood risk factor. The sample was studied at birth and then again at age 7, 11, 16 and 23 years. The outcome that Ghodsian studied was emotional disturbance at age 23, as indicated by a high score on the 'Malaise Inventory', a self-rating questionnaire that covers a range of anxiety and depressive symptoms.

The findings from both studies showed a substantial increase in the rate of impaired psychosocial functioning in early adult life associated with the specified childhood risk factor. Thus, about a third of the institution-reared subjects showed generally poor psychosocial functioning in adult life compared with just over one in ten of the men and zero per cent of the women in the comparison group. In the NCDS, women who had experienced socio-economic disadvantages in childhood had a 27% rate of emotional disturbance at 23 compared with a rate of 8% in those reared in prosperous circumstances—a three-fold difference. There was a four-fold difference for men: 12% versus 3%. We may conclude that in each case the childhood risk factor was associated with a marked increase of psychosocial problems in adult life. It will be appreciated that in both studies the risk variable reflected a heterogeneous mixture of adverse experiences that lasted over many years. It is of course important to tease out the particular facets of these experiences that created the risk, and some progress has been made in that connection; however, that issue is beyond the scope of this paper. It should be added, too, that the risk variable probably also included a genetic component; nevertheless, detailed analyses suggested that for the outcomes considered here, environmental risks predominated (Quinton & Rutter 1988, Rutter & Quinton 1987). The focus of this paper is not on these issues, but rather on the diversity of the routes by which childhood experiences may lead to impaired psychosocial functioning in adult life.

Effects on childhood behaviour that then persists into adult life

The first route involves an immediate effect leading to emotional or conduct disturbance in childhood that then persists into adult life. This mechanism was strong in both studies. Figure 1 shows the findings in the 'in care' study. Over half (56%) the institution-reared boys (compared with 29% of controls) showed conduct disturbance in childhood and in the majority of cases this was followed by pervasive social malfunction in adult life. Accordingly, 37% of cases versus 15% of controls showed conduct disorder followed by psychosocial impairment in adult life. The absolute rate of conduct disorder in females was lower but the case-control difference was even greater (19% versus 3% for conduct disturbance in childhood followed by pervasive social malfunction in early adult life).

These differences are striking, but it is clear that they underestimate the strength of continuities in emotional/behavioural disturbance between childhood and adult life (Zoccolillo et al 1991). This is both because the prospective measures in childhood did not cover the whole of the childhood age period and because measurement error has not been taken into account. When retrospective data are included and a latent class analysis is used to provide a better estimate of the underlying construct of conduct disorder, continuities are seen to be much stronger. As this finding runs counter to prevailing views on the supposed weakness of behavioural continuities between childhood and adult life, it is necessary to consider in greater detail just what the data mean. Four main points

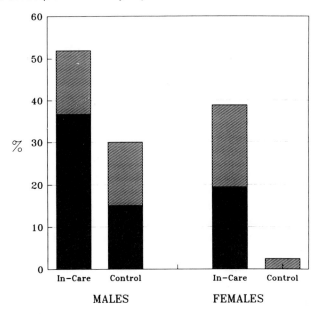

FIG. 1.   Childhood conduct disorder and adult psychosocial malfunction in institution-reared (In-Care) and control subjects. Filled columns, adult psychosocial malfunction preceded by conduct disorder. Hatched columns, adult psychosocial malfunction not preceded by conduct disorder.

need to be made. First, the continuities that we have found apply to disorders in childhood that have been persistent over prolonged time periods in early life. It is highly likely that the same continuity with adult psychopathology would not apply to the more common, transient emotional and behavioural disturbances seen in childhood. Second, the continuity particularly applies to *pervasive* social malfunction in adult life; it was not found for difficulties in single domains. Thus, most marital difficulties in otherwise well-functioning individuals were *not* preceded by conduct disorder in childhood. Third, although empirical data on this point are sparse, the continuity may be particularly great in high risk populations. Put another way, most of the increased risk for adult social malfunction or psychopathology (at least for the more common varieties) associated with adverse environments in childhood is already shown by emotional or conduct disturbances during the childhood years. The risk following adversities in childhood for psychosocial disorders developing *de novo* in adult life is much smaller. Fourth, the continuity is often heterotypic; that is, it involves a change in form over time. Thus, very few of the delinquent or conduct-disordered girls in the in-care study exhibited criminal behaviour in adult life (although the boys did do so); but the women did show many other psychosocial difficulties.

The same heterotypic continuity was evident in the NCDS data. As in other studies, the childhood risk factor was associated with a marked increase in emotional/conduct disturbance at the time. Children living in highly disadvantaged circumstances were nearly five times as likely to show marked persistent conduct disturbance as children from a prosperous background. There was a similar, although not quite so strong, association with emotional disturbance in childhood. As might be expected, emotional difficulties in childhood were associated with a two- to three-fold increase in the rate of emotional disturbance at 23 years of age (as shown by reading down the verticals in Table 1—from 3% to 10%, 5% to 11%, and 8% to 17%). However, what was perhaps surprising is that there was a similar (although not quite so great) increase in the risk of adult emotional disturbance associated with conduct disorder in childhood (as shown by reading across the horizontals in Table 1—from 3% to 8%, 6% to 8%, and 10% to 17%). This is particularly striking because the measures were based on extensive data (both parent and teacher questionnaires at 7, 11 and 16 years) and because the continuities stemming from conduct disturbance in childhood held even after controlling for emotional symptomatology. The tendency for childhood conduct disorders in girls to be followed by emotional disturbances in adult life has also been noted in other studies (Quinton et al 1990).

It is evident that the importance of effects deriving from emotional/conduct disturbances in childhood that persist into adult life is likely to have been underestimated in the past because heterotypic continuity has concealed the strength of persistence of disturbance.

Increased risk in occurrence of psychosocial hazards in adult life

The strength of psychopathological continuities between childhood and adult life is an important finding but it fails to account for the whole of the adult risk. In the first place, it begs the question of why disorders persist over such long time spans in spite of major changes in environmental conditions. Secondly, however, there are some risks for adult social malfunction associated with risk environments in childhood that are not mediated by psychopathology persisting

TABLE 1 Emotional/conduct disturbance in childhood and probability of high 'malaise' score at 23 years (males)

	Emotional disturbance		
	Low	Medium	High
Conduct disturbance			
Low	0.03	0.05	0.08
Medium	0.06	0.07	0.08
High	0.10	0.11	0.17

National Child Development Study; M. Ghodsian, unpublished data 1989.

from childhood. A mechanism that appears relevant for both issues is the generation of risk environments in adult life.

This is apparent, for example, in the NCDS finding of a nearly three-fold increase in the rate of socio-economic disadvantage at 23 years associated with marked conduct disturbance in childhood compared with no conduct problems (Table 2, 35% versus 13%)—a difference that remained significant after taking account of socio-economic disadvantage in childhood. This association has been evident in previous longitudinal studies from Robins' (1966) pioneering follow-up study onwards. It is clear that individuals showing aggressive or antisocial behaviour act in ways that tend to put them in psychosocially risky environments.

However, this tendency is by no means confined to socially disruptive behaviour. Thus, in the in-care study, the women reared in institutions were much more likely than controls to become teenage parents (41% versus 5% in controls) and to marry (or cohabit with) a deviant partner (52% versus 9%)—meaning someone who exhibited psychiatric disorder, delinquent behaviour, drink or drug problems, or longstanding difficulties in interpersonal relationships. The differences for men were much smaller and statistically non-significant (27% versus 17% and 8.9% versus 2.4% respectively; see Fig. 2). The consequence was that the institution-reared women were much more likely than controls to lack marital support at follow-up. This difference was much less apparent in men. However, in both men and women, marital support when present was associated with a quite powerful protective effect (Fig. 3).

Obviously, it was important to consider the possibility that this association represents an artefact consequent upon the individuals' own deviant functioning. However, rigorous analyses failed to support that suggestion (Pickles & Rutter 1991, Quinton & Rutter 1988). There was a real turning point in life trajectory by which a harmonious marriage to a non-deviant spouse significantly reduced the risk of adult social malfunction in females who had shown conduct disorder in childhood. There was some tendency for deviant girls to choose a deviant man as partner, but the effect was weak in comparison with the strength of

TABLE 2   Conduct disturbance in childhood and socio-economic disadvantage at 23 years (males)

| | Conduct disturbance | | | |
	None (%)	Some (%)	Moderate (%)	High (%)
Socio-economic disadvantage at 23 years				
Low	56	50	43	34
Medium	32	31	31	32
High	13	20	27	35

National Child Development Study; M. Ghodsian, unpublished data 1989.

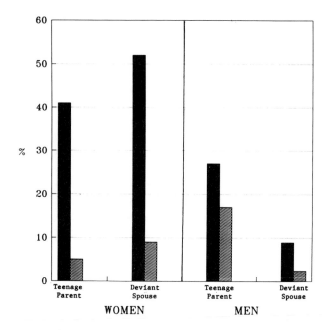

FIG. 2. Teenage parenthood and deviant first spouse in institution-reared (filled columns) and control (hatched columns) men and women.

effect associated with the mediating variable of 'planning'—meaning a tendency to exert definite choice in jobs or careers, and marriage. Women who exerted planning in their lives were some five times more likely to have marital support at follow-up. The tendency in men was similar but much weaker. This sex difference was probably a function of the fact that institution-reared women were very likely to cohabit at age 18 or younger (40% versus 5% in controls), whereas this was less the case with the men (18% versus 12%).

An event history model applied to the in-care women (Pickles & Rutter 1991) showed that cohabitations at younger ages were much more at risk of being with a deviant spouse than those at older ages (Fig. 4). The tendency in our culture for men (even those from a high risk background) to marry later than women was probably a protective factor for them with respect to the hazard of marriage to a deviant spouse. This did not operate in women, but the characteristic of planning or not planning (related in part to whether or not they had good experiences at school outside the institutional environment—see Quinton & Rutter 1988) played a key mediating role in determining the quality of the marital relationship which, in turn, had a substantial effect on that person's functioning in other social domains.

These findings underline the importance of chain effects by which early environmental risk experiences lead to a behavioural style (*not* strongly

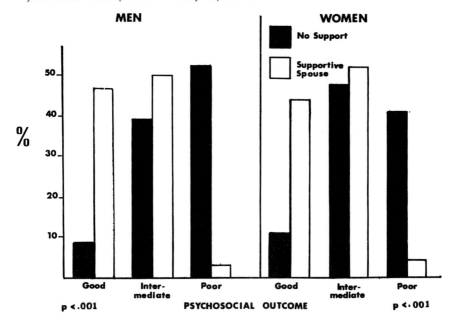

FIG. 3. Marital support and social functioning in men and women after institutional rearing. Filled columns, no support; unfilled columns; supportive spouse.

associated with psychopathology) that, in conjunction with cultural influences, leads further to a risky environment in adult life that predisposes to psychosocial malfunction in adult life.

Increased vulnerability to psychosocial hazards in adult life

The route just considered concerned an increased likelihood of *experiencing* adverse environments. A further possible mechanism concerns an increased *vulnerability* to such environments. We examined that possibility with respect to the effects of institutional rearing on parenting in relation to susceptibility to the effects of lack of marital support in adult life. The suggestion from the findings (Fig. 5) is that there is a tendency for the institution-reared women to be more likely to exhibit difficulties in parenting when faced with marital problems. The interaction fell just short of statistical significance, key cell sizes being small, and the possibility requires further study in a larger sample before firm conclusions are drawn. However, taken in conjunction with other data from the same study (for example, the much greater tendency of the institution-reared women to give up the care of their children when faced with serious family difficulties—see Quinton & Rutter 1988) and from other investigations (e.g. the already noted vulnerability effect in Rodgers' 1990 study), it is clear that the

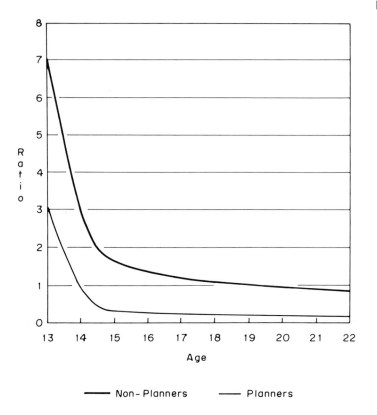

FIG. 4.   Ratio of hazard of marriage to deviant and non-deviant spouse in 'planners' and 'non-planners' in relation to age at marriage.

mechanism of increased vulnerability to psychosocial hazards, perhaps as a result of less adequate coping skills, is a plausible one that warrants more systematic testing.

Conclusions

In summary, the data from two long-term longitudinal studies, dealing with the risks associated with rearing in an institution or in socially disadvantaged circumstances, both show substantial continuities between childhood experiences and psychosocial functioning in adult life. To an important extent, these long-term risks were already manifest in childhood through a contemporaneous effect on emotional or conduct disturbance that then persisted into adult life. Such psychopathological continuities are stronger than was usually supposed in the past, their strength having been concealed by a change in the form of the disturbance. However, the findings have also indicated that the long-term risk

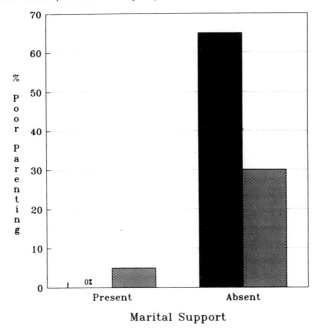

FIG. 5.   Status of current parenting (expressed as % poor parenting, ordinate) in relation to the presence or absence of marital support in institution-reared (filled columns) and control (hatched columns) subjects.

is mediated in part through an effect on the adult environment, the risk of a hazardous environment in adult life having been increased by styles of behaviour (not involving overt psychopathology) that serve to shape or select later environments in ways that increase psychosocial risk. In addition, although less well documented, it may well be that adverse experiences in childhood impair coping skills and adaptive resources so that there is an increased vulnerability to adverse environments experienced in adult life.

Acknowledgements

I am most grateful to Mayer Ghodsian and David Quinton for permission to use data from their respective studies. The former derive from a study undertaken on a grant from the Economic and Social Research Council to Mayer Ghodsian and Stephen Wolkind.

References

Clarke AM, Clarke ADB 1976 Early experience: myth and evidence. Open Books, London
Pickles A, Rutter M 1991 Statistical and conceptual models of 'turning points' in developmental processes. In: Magnusson D, Bergman LR, Rudinger G, Törestad B (eds)

Problems and methods in longitudinal research: stability and change. Cambridge University Press, Cambridge, in press

Quinton D, Rutter M (eds) 1988 Parenting breakdown: the making and breaking of inter-generational links. Avebury, Aldershot

Quinton D, Rutter M, Gulliver L 1990 Continuities in psychiatric disorders from childhood to adulthood in the children of psychiatric patients. In: Robins L, Rutter M (eds) Straight and devious pathways from childhood to adulthood. Cambridge University Press, Cambridge, p 259–278

Robins L, Rutter M (eds) 1990 Straight and devious pathways from childhood to adulthood. Cambridge University Press, Cambridge

Robins LN (ed) 1966 Deviant children grown up. Williams & Wilkins, Baltimore

Rodger B 1990 Influences of early-life and recent factors on affective disorder in women: an exploration of vulnerability models. In: Robins L, Rutter M (eds) Straight and devious pathways from childhood to adulthood. Cambridge University Press, Cambridge, p 314–327

Rutter M (ed) 1981 Maternal deprivation reassessed. Penguin, Harmondsworth, Middlesex

Rutter M 1987 Continuities and discontinuities from infancy. In: Osofsky J (ed) Handbook of infant development. Wiley, New York, 2nd edn, p 1256–1296

Rutter M 1989 Pathways from childhood to adult life. J Child Psychol Psychiatry Allied Discip 30:23–51

Rutter M, Pickles A 1991 Person–environment interactions: concepts, mechanisms and implications for data analysis. In: Wachs TD, Plomin R (eds) Conceptualization and measurement of organism–environment interaction. American Psychological Association, Washington, DC, in press

Rutter M, Quinton D 1987 Parental mental illness as a risk factor for psychiatric disorders in childhood. In: Magnusson D, Ohman A (eds) Psychopathology: an international perspective. Academic Press, New York, p 199–219

Rutter M, Quinton D, Hill J 1990 Adult outcome of institution-reared children. In: Robins L, Rutter M (eds) Straight and devious pathways from childhood to adulthood. Cambridge University Press, Cambridge, p 135–157

Scarr S, McCartney K 1983 How people make their own environments: a theory of genotype–environment effects. Child Dev 54:424–435

Zoccolillo M, Pickles A, Quinton D, Rutter M 1991 The outcome of conduct disorder: implications for defining adult personality disorder. Submitted

DISCUSSION

Wood: In clinical practice we often see disturbed and ill children who come from one-parent families, and a rather provocative political statement has been made in the UK that such families are 'wrecking' society. Do you think that the present rather high number of one-parent families has its origins in past hardships experienced over two generations, including especially wartime separation for the grandparental generation because of the evacuation of children from the cities, which was a significant experience for many urban populations in Britain? Alternatively, could it be that aspects of the social climate more recently may have reduced, albeit at superficial level, only the disincentives to being a one-parent family?

Rutter: A bit of each! Unquestionably there has been a marked UK increase both in one-parent families and in children born out of wedlock (Wicks & Kiernan 1990), but it would be mistaken to assume that that means the children were unplanned and unwanted. More single women are choosing to have children and to bring them up on their own. Similarly, more couples are choosing to have children but yet not to marry. These decisions are associated with costs, but there is now a greater acceptance of non-traditional family patterns and we know very little about the risks to the children involved in these changed circumstances.

The reasons for the increased rate of marital breakdown are also difficult to discern (Rutter 1979/80). It does not seem likely that experiences during the Second World War were critical. That was a popular explanation for delinquency at one time, but the timing of associations proved to be wrong. However, there is evidence that patterns of upbringing affect people's behaviour as parents, as my paper indicated.

Wadsworth: It is important to differentiate those who experienced parental divorce or separation in childhood from those who experienced parental death, or no such change in the family. In the 1946 birth cohort, those who had experienced parental divorce were then significantly more likely than others to experience the chains of risk of the kind you describe. They were more likely than others to experience relative loss of socio-economic status in childhood and to move schools, to have much less likelihood of attainment in further or higher education (Wadsworth & Maclean 1986), and consequently to have reduced chances of high income or socio-economic status in their own adult life (Maclean & Wadsworth 1988). In women in adulthood there were also significantly greater risks of emotional disturbance and high rates of alcohol consumption among those who had, in childhood, experienced parental divorce or separation (Kuh & Maclean 1990).

Barker: Can you tell us what you hope to learn about the children of men and women in the 1946 birth cohort?

Wadsworth: We have studied the first-born offspring of the 1946 birth cohort at the ages of four and eight years. So far we have shown that parental education and socio-economic status are important for the educational chances of offspring. Children of parents with poor educational attainment and of low socio-economic status were the least likely to have had any kind of preschool or kindergarten experience by four years, and those without such experience were relatively poor scorers on verbal attainment tests taken at age eight years (Wadsworth 1981).

Richards: The choice of marriage partner may be an important mediator of cross-generational effects. This is something that interests those of us who look at the effects of parental divorce. In the UK, various changes in the benefit system and taxation seem likely to affect the time when young people leave home. Young people may be tending to leave their parental home earlier. Leaving home earlier is known to follow parental divorce. If that effect is accentuated, young people may be more likely to enter into relationships that may not be supportive, even earlier than they do now. To look at this in a more cross-cultural way, this may be a particular problem of Northern and Western Europe and North America. In these regions, young people select their own partners, but in fact this is rather an odd thing to do. In most cultures of the world, your parents and family choose your marital partner, and very often they may select someone in your family, because that can maintain economic links.

These differences in marriage systems may be very important from the point of view of the long-term effects we are discussing. Our marriage system, based on early individual choice, may make the transmission of adversity across generations more likely. I think one can see this happening. This is the picture with which we are familiar, with many single-parent households, some very young mothers and a concentration of poverty in lower social classes that tends to be transmitted across generations. I would suggest that this same pattern of individual adversity and poverty may not be transmitted across generations where arranged marriages are the norm and where the wider family is much more involved in family life. These societies may break the transmission because marriage is not a matter of individual choice for young people, and the wider family protects against individual adversity. I suggest that our Northern European pattern of marriage has a lot to do with the intergenerational transfer of adversity that you demonstrate so beautifully.

Rutter: It would be very interesting to have systematic cross-cultural comparisons. Are there systematic studies of the effects of qualities of the marital partner in societies with arranged marriages?

Richards: We know very little about this, so we are speculating from very limited data. And there are confounding factors. In societies where marriage is arranged, the broader kinship network tends to offer a lot of support to young couples and their children, and therefore the qualities of a spouse may be much less important than in our society.

Rutter: There are large national differences too, in teenage pregnancy; the US stands out as different from the rest of the world in having much the highest rate, followed by Canada, with European countries way below (Hayes 1987).

Murray: It must be very difficult to work out the direction of the effects involving supportive spouses. How do you know that as the outcome, by chance, gets better, a spouse may not *become* more supportive?

Rutter: Our data analyses were designed to examine just that issue. Of course, there must be two-way interactions, but still the evidence indicates the strong likelihood of a true supportive effect (Quinton & Rutter 1988, Pickles & Rutter 1991). There are several pieces of evidence that point in that direction. For example, the supportive effect was evident on the basis of the spouse's characteristics *before* marriage or cohabitation. Also, the effect remained after control for the girl's own prior behaviour and for measurement error of that behaviour. Thus, antisocial girls in the study were rather unlikely to land up with a supportive non-deviant partner, but, in the instances where they did, there was an effect of marital support leading to better social functioning.

Suomi: It is a joy, Dr Rutter, to hear these basic principles that seem to apply across different primate species as well as cross-culturally, particularly your notion of single measures of symptoms underestimating the continuity of specific disorders, and the resulting heterotypic nature of most forms of disturbance, especially from the developmental standpoint. This means that expressions of behavioural problems will be very different at different ages or different degrees of maturational status, and one must take that into account when trying to understand issues of continuity. Also, I applaud your emphasis on the power of the current environment, however you wish to characterize it, in terms of both external forces and resources such as spouse support. And finally, one should recognize the fact that males and females do not have the same prognosis, given comparable starting points, and one shouldn't always expect the same mechanisms to hold for males and females.

Given all of this, one might care to examine the reverse situation and consider instances of good outcomes, or examples of individuals who as adults are in a state of well-being, despite having some of these early problems. Can these data sets be used to identify possible routes whereby favourable outcomes emerge, and might they give us ideas about intervention strategies?

Moxon: I have a similar point. You have noted the negative sequelae following early psychosocial adversities, but your data also show that many people subjected to these experiences were functioning quite well at follow-up, indeed as well as the control group. Is it not important also to focus on the factors associated with good outcomes in spite of early adversities?

Rutter: I entirely agree with both of you. We have had a particular interest in the study of resilience and in the factors involved in protective mechanisms (Rutter 1985, 1987) and some leads are available. For example, we sought to determine why some young people with an institutional upbringing exerted planning in their lives, whereas others did not. It turned out that positive experiences at school were particularly important. The Local Authority policy at that time, during the 1960s, meant that children from the two Group Homes that we studied were distributed across a wide range of primary and secondary schools (the distribution being designed to avoid their being labelled as institutional children). The effect of the policy was to create an unusual disjunction between home and school environments, allowing the latter to have a greater than average effect. Good experiences there—whether in sport, music, crafts, positions of responsibility or academic work—were associated with a greater tendency to exert planning. It seemed that success in one arena could lead to a more effective style of coping that carried benefits in other situations. It also appeared that institutional policies designed to take decision-making away from individuals (a feature of most institutions) may well be damaging. People need to be able to learn from making decisions and taking responsibility, and that must include the chance of making mistakes. Sometimes there is a tendency to define protective factors just in terms of happy experiences. The analogy with protection from infections is perhaps informative. It is not positive healthy experiences that are protective but, rather, controlled exposure to the pathogens in circumstances that allow the body to cope successfully—the basis of immunization and also natural immunity. It appears that the same may apply in the psychological arena. Probably, resilience does not mainly derive from things that make you happy, but rather from the self-esteem and coping skills that come from successful encounters with earlier challenges and stressors.

Barker: Are you telling us something new when you talk about these 'heterotypic' associations, Professor Rutter? Our follow-up study of middle-aged men in Hertfordshire shows essentially that a set of adverse influences in very early life not only increase your risk of entering a coronary care unit, but reduce your likelihood of being selected for the Olympic Games, by virtue of shorter stature! In a sense, those are 'heterotypic' or unexpected associations. Are you not simply saying that you are rather surprised that there are mechanisms which are more complicated than you first thought?

Rutter: By heterotypic continuity I do not mean unexpected associations, nor do I mean varied outcomes. Rather, the point is that the same underlying variable may manifest itself in different ways at different points in development—a common feature of the developmental process. For example, the propensity to form close reciprocal relationships is a feature of human beings at all ages after early infancy, but obviously the relationships formed at the age of one year look rather different from those at, say, 16 years, or in old age. Even more strikingly, it has been shown that the infancy equivalent of general intelligence is visual attention and not the developmental accomplishments that have a greater surface similarity to intelligence (Bornstein & Sigman 1986). Of course, before it is assumed that two apparently different behaviours reflect the same underlying construct, there is an obligation to show that they are functionally equivalent. The mere demonstration of a significant correlation over time between them is not enough.

Barker: When you know more about psychopathological mechanisms, perhaps you will drop the word 'heterotypic' and you will say that there are two sorts of mechanisms: there are the immediately recognizable consequences of adversity which persist, and there are consequences which increase vulnerability to later challenges. These are two concepts that we have had reiterated here for physical illness.

Rutter: I agree that one of the mechanisms underlying continuity is a vulnerability to later challenges, but that is not what is meant by heterotypic continuity.

Casaer: In developmental neurology we now realize that to look at the same variables at different ages is not always the optimal thing to do, if one's interest is the understanding of brain development. It seems more appropriate to search for age-specific behavioural sets that can be linked to underlying maturational changes in the brain. During this meeting we have mentioned a few examples: feeding and breathing in the neonate, visual smiling at about eight weeks, and the interaction between postural and fine motor control at the end of the first year of life. What is your approach in the study of behavioural development within a developmental neuropsychiatric frame of reference?

Rutter: The approach that you suggest is certainly applicable, but, at least with respect to psychopathology, the measures of key predictive value may not necessarily be those that are most pertinent to the developmental issues at that age. For example, to go back to Robin Murray's paper, it is clear that roughly half the individuals who develop schizophrenic psychoses have shown substantial behavioural abnormalities in childhood (Rutter & Garmezy 1983). These abnormalities (which involve attentional deficits, social abnormalities, and disruptive behaviour) do not look like schizophrenia (thus, they do not include delusions, hallucinations, thought disorder and the like), but equally they do not represent behaviours of particular developmental relevance either. Instead, probably they represent the underlying genetically determined neuro-developmental disorder.

On the other hand, the point that you make might apply to the developmental predictors of reading retardation. Satz (Satz & van Nostrand 1979) suggested that children who are poor readers will have had visuomotor problems in early childhood, but language difficulties in middle childhood. The available evidence does not support the concept at all clearly (Satz et al 1976), but the key test is whether these contrasting deficits are shown by the *same* individuals at different ages (the alternative is that both deficits may predispose to reading difficulties, but that they do so in different children through different mechanisms).

Lucas: You have talked about early influences on later social outcome, but pointed out that there are many knock-on effects and great difficulties in interpretation because of interactions with later events. Is there a parallel here with what we have been discussing in biological terms, of a sensitive period? Are there periods when certain aspects of social and psychological development take place in a child which can't be deferred to another time? And, if so, is there always a permanent distortion of psychosocial development, if they haven't taken place at the right time?

Rutter: The answer depends on which psychological characteristics are being considered. Although the notion of fixed, narrow critical periods rightly came under severe criticism in the 1970s (having been fashionable in the 1950s and 1960s), it may well be that there are age-dependent sensitivities to particular sorts of experiences. For example, Hodges & Tizard (1989a,b), in their follow-up to age 16 of children from a residential nursery, contrasted those who were restored to their biological families (most of which provided a rather unsatisfactory environment) and those who were adopted at about three or four years into well-functioning families. Antisocial behaviour was largely a function of the environment at follow-up (being much commoner in the restored group) but the quality of social relationship seemed to be a function of the environment in the first few years. Both groups (who shared the infancy experience of rearing in a residential nursery) differed from controls in being less likely to confide and being less likely to make intense peer relationships. The data are open to more than one interpretation, but the findings seem to suggest some sort of sensitive period effect for the development of the social qualities needed for close peer relationships in adolescence.

Wadsworth: This point about what is measured and at which stage in life is important. In looking for later risk factors from an earlier life problem it is necessary to examine a wide range of possible outcome measures, because the expression of continuing vulnerability may take a number of forms, including problem behaviour and illness of suspected psychosomatic origin (Wadsworth 1984). The form of expression taken is likely to be affected not only by the nature of the earlier life problem, but also by the chain of risk experienced, by personality and by genetic predisposition. The expression of continuing vulnerability will also be related to life stage, because there are peak age periods for many of the indicators used, such as delinquent behaviour, criminality in men and depression in women.

Martyn: You compared your ex-care group and the control group with regard to current performance as parents, with different levels of marital support. There appeared to be evidence that the group that received marital support and had been in care were better as parents than the comparison group (Fig. 5, p 199). It may be that you are not very impressed by the size of the difference, but it is a rather unusual example of a bad experience in the first part of life working out, if not to one's own advantage, at least to that of one's offspring!

Rutter: In fact, those were statistically non-significant differences. Nevertheless, the notion that the same bad experience may have a sensitizing effect in some individuals but a steeling effect in others is a plausible one, well worth further study.

Richards: There's an old idea in the literature that the question of choice of partner may be bimodal after a parental divorce. It can make you very cautious about marriage and very careful in your choice of partner, and make you postpone marriage, or it can lead to an early marriage after a very brief courtship—and a marriage that may be forced by a pregnancy.

Wadsworth: We have found that one way in which an early bad experience may lead to a potentially good outcome is after the death of a parent. In the 1946 birth cohort, children who experienced parental death tended to do much better than expected in their subsequent educational attainment when compared not only with children who experienced other forms of parental loss, such as divorce or separation, but also in comparison with children who did not experience any change in family structure (Wadsworth & Maclean 1986).

Hanson: Professor Rutter, you mentioned ways of assessing the degree of error in the measurement of your variables. How is that done?

Rutter: My way is to rely on colleagues who know more about it than I do—in particular my colleague, Andrew Pickles. The basic approach is to use mathematical techniques, such as structural equation modelling, to derive the underlying construct that each of the measures is designed to tap (Fergusson & Horwood 1988, Pickles & Rutter 1991, Rutter & Pickles 1991). A questionnaire score on, say, aggression, will necessarily provide an imperfect, and biased, measure of the hypothesized trait of pervasive aggressivity. Each score will reflect a mixture, consisting of something that reflects that trait, a reflection of other traits that happen to be associated with aggression, situational influences, rating biases, and random error. Multiple measures, preferably over several time points, enable one to separate out these different elements. The intercorrelations between the measured variables, together with their base rates in high and low risk groups, can be used mathematically to provide error-free estimates of the underlying hypothesized construct. The approach requires multiple data points but its use can be extremely helpful in the study of continuities and discontinuities over time.

References

Bornstein MH, Sigman M 1986 Continuity in mental development from infancy. Child Dev 57:251–274

Fergusson DM, Horwood LJ 1988 Structural equation modelling of measurement processes in longitudinal data. In: Rutter M (ed) Studies of psychosocial risk: the power of longitudinal data. The European Science Foundation. Cambridge University Press, Cambridge

Hayes CD 1987 Preventing adolescent pregnancy: an agenda for America. National Academy Press, Washington, DC

Hodges J, Tizard B 1989a IQ and behavioural adjustment of ex-institutional adolescents. J Child Psychol Psychiatry Allied Discip 30:53–75

Hodges J, Tizard B 1989b Social and family relationships of ex-institutional adolescents. J Child Psychol Psychiatry Allied Discip 30:77–79

Kuh D, Maclean M 1990 Women's childhood experience of parental separation and their subsequent health and socio-economic status in adulthood. J Biosoc Sci 22:121–135

Maclean M, Wadsworth MEJ 1988 The interests of children after parental divorce: a longterm perspective. Int J Law & Family 2:155–166

Pickles A, Rutter M 1991 Statistical and conceptual models of 'turning points' in developmental processes. In: Magnusson D, Bergman LR, Rudinger G, Törestad B (eds) Problems and methods in longitudinal research: stability and change. Cambridge University Press, Cambridge, in press

Quinton D, Rutter M (eds) 1988 Parenting breakdown: the making and breaking of intergenerational links. Avebury, Aldershot

Rutter M (ed) 1979/80 Changing youth in a changing society: patterns of adolescent development and disorder. Nuffield Provincial Hospitals Trust (1979), Harvard University Press Boston, MA (1980)

Rutter M 1985 Resilience in the face of adversity: protective factors and resistance to psychiatric disorder. Br J Psychiatry 147:598–611

Rutter M 1987 Psychosocial resilience and protective mechanisms. Am J Orthopsychiatry 57:316–331

Rutter M, Garmezy N 1983 Developmental psychopathology. In: Hetherington EM (ed) Socialization, personality, and social development. Handbook of child psychology, vol 4 (4th edn). Wiley, New York p 775–911

Rutter M, Pickles A 1991 Improving the quality of psychiatric data: classification, cause and course. In: Magnusson D, Bergman LR (eds) Data quality in longitudinal research. Cambridge University Press, Cambridge, p 32–57

Satz P, van Nostrand GK 1979 Developmental dyslexia: an evaluation of a theory. In: Satz P, Ross JJ (eds) The disabled learner: early detection and intervention. Rotterdam University Press, Rotterdam

Satz P, Friel J, Rudegeair F 1976 Some predictive antecedents of specific reading disability: a two, three and four year follow-up. In: Guthrie JT (ed) Aspects of reading acquisition. Johns Hopkins Press, Baltimore

Wadsworth MEJ 1981 Social class and generation differences in pre-school education. Br J Sociol 32:560–582

Wadsworth MEJ 1984 Early stress and associations with adult health behaviour and parenting. In: Butler NR, Corner BD (eds) Stress and disability in childhood. John Wright, Bristol, p 100

Wadsworth MEJ, Maclean M 1986 Parents' divorce and children's life chances. Children & Youth Services Rev 8:145–159

Wicks M, Kiernan K 1990 Family change and future policy. Family Policy Studies Centre/Joseph Rowntree Memorial Trust, London

Prolegomena to a model of continuity and change in behavioural development

Avshalom Caspi

Department of Psychology, 1202 West Johnson Street, University of Wisconsin, Madison, WI 53706, USA

Abstract. It is now widely acknowledged that personality and behaviour are shaped in large measure by interactions between the person and the environment. There are many kinds of interaction but we suggest that three types play particularly important roles both in sustaining behavioural continuity across the life course and in guiding the trajectory of the life course itself. *Reactive* interaction occurs when different individuals exposed to the same environment experience it, interpret it and react to it differently. *Evocative* interaction occurs when an individual's personality evokes distinctive responses from others. *Proactive* interaction occurs when individuals select or create environments of their own. Within this framework we also examine systematic change and turning points in behavioural development. We have recently advanced a paradoxical theory suggesting that behavioural continuities are especially likely to be evident during periods of social discontinuity; that is, dispositional factors influence behaviour most when individuals enter new situations and assume new statuses. This model receives empirical support from both experimental and longitudinal-correlational research. The model also presents interesting implications for our understanding of turning points in behavioural development: to effect change in the life course, new situations must eliminate old options and create new opportunities. Convergent evidence from experimental and naturalistic designs is introduced to support this claim.

1991 The childhood environment and adult disease. Wiley, Chichester (Ciba Foundation Symposium 156) p 209–223

What processes or mechanisms promote, enhance, and disrupt continuities across the life course? My purpose here is to highlight some of the conceptual issues involved in answering these questions.

What promotes continuity?

Longitudinal studies have repeatedly confirmed that individual differences in personality characteristics are stable across time and circumstance. Efforts to go beyond description to the more difficult task of explanation are, however, far less developed. How is continuity possible amid the myriad of social changes and transformations that characterize a human life? Recent attempts to integrate interactional processes of personality functioning may offer some answers about the sources of behavioural continuity (Caspi & Bem 1990).

Person–environment interactions

There are many kinds of interaction (Rutter 1983), but we suggest that three types play particularly important roles in sustaining behavioural continuity across the life course (Buss 1987, Plomin et al 1977, Scarr & McCartney 1983).

Reactive interaction occurs when different individuals exposed to the same environment experience it, interpret it, and react to it differently. Each individual extracts a subjective psychological environment from the objective surroundings, and that subjective environment is what shapes both personality and subsequent commerce.

This is the basic tenet of the phenomenological approach historically favoured by social psychology and embodied in the dictum that if people 'define situations as real, they are real in their consequences' (Thomas & Thomas 1928). It is also the assumption that connects Epstein's (1980) writing on the development of 'personal theories'; Tomkins' (1986) description of 'scripts' about the self and interpersonal interactions; and Bowlby's (1973) analysis of 'working models'— mental representations of the self and others—that develop in the context of interactional experiences.

Why should personal theories, scripts and working models promote continuity? One possibility is that these cognitive-dynamic structures function as filters for social information. In fact, social psychologists, who tend to focus on the cognitive rather than the motivational features of internal organizational structures, argue that 'self-schemas'—psychological constructs of the self— screen and select from experience to maintain structural equilibrium (Greenwald 1980). Once a schema becomes well organized, it filters experience and makes people responsive to information that matches their expectations and views of themselves and others (Markus 1977). Cantor & Kihlstrom (1987) review a host of cognitive processes that may promote consistency and impair people's ability to change.

As I shall indicate next, however, the use of existing structures to organize experience is not entirely unconscious. Persistent ways of perceiving, thinking and behaving are not preserved simply by psychic forces, nor are they entirely attributable to features of the cognitive system; they are also maintained by the consequences of everyday action (Wachtel 1977).

Evocative interaction occurs when an individual's personality evokes distinctive responses from others. The person acts, the environment reacts, and the person reacts back in mutually interlocking evocative interaction.

The ways in which evocative interaction can promote continuity has been elegantly shown by Patterson's (1982) work with aggressive boys. The boy's coercive behaviours initiate a cycle of parental anger and further aggression until the parents finally withdraw, thereby reinforcing the boy's initial aggression: 'Family members and antisocial children alternate in the roles of aggressors and victims, each inadvertently reinforcing the coercive behavior of the other' (Patterson & Bank 1990). In interpreting longitudinal data, we are often alerted to the possibility that continuities observed in trait measures of maladaptive behaviour may simply reflect the cumulative and continuing continuities of noxious environmental influences. Patterson's work shows that behavioural patterns can themselves create such stable environments.

By extension, we have suggested that a child whose temper tantrums coerce others into providing such short-term payoffs in the immediate situation may learn an interactional style that continues to work in later years. The immediate reinforcement not only short-circuits the learning of more controlled interactional styles that might have greater adaptability in the long run; it also increases the likelihood that coercive behaviours will recur whenever similar interactional conditions arise later in the life course (Caspi et al 1989).

It is also through evocative interaction that phenomenological interpretations of situations—the products of reactive interaction—are transformed into situations that are 'real in their consequences'. Early experiences can set up expectations that lead an individual to project particular interpretations onto new situations and relationships and thence to behave in ways that corroborate those expectations (Wachtel 1977). For example, because aggressive children expect others to be hostile (Dodge 1986), they may behave in ways that elicit hostility from others, thereby confirming their initial suspicion and sustaining their aggression.

Individuals also elicit and selectively attend to information that confirms rather than disconfirms their self-concepts (Darley & Fazio 1980, Snyder 1984). This promotes stability of the self-concept which in turn promotes the continuity of behavioural patterns that are congruent with that self-concept (Snyder & Ickes 1985).

In these several ways, then, reactive and evocative interactions enable an ensemble of behaviours and expectations to evoke maintaining responses from others—thereby promoting continuity across time.

Proactive interaction occurs when individuals select or create environments of their own. These environments, in turn, can promote the continuity of personality. Scarr & McCartney (1983) have proposed that this dispositionally guided selection and creation of environments becomes increasingly influential in development as the child gains increased autonomy.

The environments that are most consequential for behavioural development are probably our interpersonal environments, and it is in friendship formation and mate selection that the personality-sustaining effects of proactive interaction are most intriguing.

For example, aggressive children, although unpopular in the larger community of peers, are closely linked to particular subgroups of children and tend to affiliate with friends who match their aggressive behaviour (Cairns et al 1988). According to Patterson & Bank (1990), children shop for settings and people that maximize their positive payoffs. The trial-and-error process of shopping and being rejected inevitably leads problem children to identify groups of peers that will reinforce their behaviours. Thus, the social network, whose composition is in part determined by direct social preferences, serves as a convoy through development, producing supports for acts of increasing deviance.

Research on marriage similarly indicates that people tend to choose partners who are similar to themselves (Buss 1985). Assortative mating has both genetic and social consequences (Thiessen & Gregg 1980), and it also appears to have important implications for behavioural continuity. In two longitudinal studies of married couples, Caspi & Herbener (1990) showed that marriage to a similar other is significantly related to intra-individual consistency in the organization of personality attributes across adulthood. By choosing situations that are compatible with their dispositions and by affiliating with similar others, individuals set in motion processes of social interchange that sustain their dispositions across time and circumstance.

Thus far, we have noted that early individual differences have lawful implications for the course of later behaviour and development. We have also elucidated some of the processes that may promote behavioural continuity. But the question remains: *when* are the manifestations of early individual differences most likely to emerge later in the life course? It is necessary to tackle this question before attempting to clarify processes of change.

The paradox of continuity

We have recently advanced a paradoxical theory suggesting that characterological continuities are most likely to be evident during periods of social discontinuity; that is, dispositional factors influence behaviour most when individuals enter new situations and assume new statuses. This is paradoxical because most theoretical perspectives on life discontinuities assume that these are times when major reorganizations occur. Nevertheless, our model receives empirical support from both experimental and longitudinal-correlational research.

Social psychologists have long known that situations differ from one another in the degree to which they permit individual differences to manifest themselves. Some situations are weak, permitting a variety of responses and hence a variety of individual differences to flourish; other situations are strong, constraining behavioural choices and eliciting similar responses from most individuals (Mischel 1977). In general, the influence of dispositional factors on behaviour is most pronounced in weak situations. For example, individual differences in introversion–extraversion predict behaviour better in situations that encourage neither one nor the other (Monson et al 1982). Likewise, marked dispositional influences on social behaviours emerge more strongly in unstructured dyadic interactions than in more traditional, structured experimental paradigms (Ickes 1982).

Similar findings emerge from research in behavioural genetics, where twin studies have shown that genetic effects are most pronounced in unstructured situations and in response to novel social encounters. For example, differences between monozygotic and dizygotic twin pairs are stronger in playroom settings, where the range of behavioural reactions is less restricted, than in test-room settings that are highly structured (Matheny & Dolan 1975). Differences between monozygotic and dizygotic twins are also stronger when children are confronted by unfamiliar rather than by familiar persons (Plomin & Rowe 1979).

In addition, research in developmental psychology suggests that individual differences emerge strongly in novel settings that require individuals to master and negotiate new demands and tasks. For example, Wright & Mischel (1987) have shown that the social behaviours of children diverged as the competency requirements of the situations in which they were engaged increased. More generally, they suggest that demanding and stressful situations are likely to increase response variability between individuals and may also elicit stable individual differences.

Fuller (1967) arrived at a similar conclusion on the basis of his experiential deprivation experiments. According to his emergence-stress theory, genetic/breed differences in behavioural responses are most pronounced when animals *emerge* into a complex and novel environment. More generally, ethologists have long known that animals will revert to species-typical innate behaviours when under stress and, more importantly, in situations where ambiguity is 'created by the conflict between two or more behavioral tendencies' (Wilson 1975, p 225). Novel and stressful settings often elicit entrenched responses.

Accordingly, we proposed recently that individual differences should be accentuated when individuals experience profound discontinuities (Caspi & Moffitt 1991). To test this hypothesis we studied the behavioural responses of adolescent girls to the onset of menarche in a longitudinal study of an unselected birth cohort. We first tested predictions from three rival hypotheses about the relation between pubertal change and social-psychological change: the 'stressful change', 'off-time', and 'early timing' hypotheses. The results supported the early timing hypothesis: behavioural problems were associated with puberty only when menarche occurred at a very young age. Next we sought to determine whether stressful, early menarche generated new behavioural problems or accentuated pre-menarcheal dispositions. The results supported an accentuation model: the early onset of menarche accentuated behavioural problems only among girls who were predisposed to behavioural problems earlier in childhood.

Stressful transition events, such as the early onset of menarche, do not generate uniform reactions among individuals; they appear, rather, to accentuate pre-transition differences between them. Although most theoretical perspectives on life discontinuities suggest that transitions are times when individuals are most likely to change, our data suggest that they may actually be times when pre-existing individual differences are most likely to be accentuated.

Research on other transitions in the life course similarly indicates that differences between individuals may be magnified when they move into new situations. For example, the accentuation of individual differences has been observed in children going off to school for the first time (Entwisle & Alexander 1988) as well as in students entering college (Feldman & Newcomb 1969). The accentuation of individual differences is especially apparent during stressful crises. For example, irritable and explosive men tend to become even more so during periods of severe economic setbacks (Elder & Caspi 1988). The same point is highlighted in studies of community disasters. For example, the coping styles of entrepreneurs whose businesses suffered extensive damage during a natural disaster were accentuated during the recovery periods (Anderson 1977).

In all, it appears that individual differences may be most pronounced when individuals experience profound discontinuities in their lives; that is, during transition events in the life course. Transition events that are characterized by ambiguity, novelty and uncertainty (e.g., the early onset of menarche) are weak situations to the extent that they allow each individual to encode and experience the discontinuity in a non-uniform and idiosyncratic way (Mischel 1977).

More generally, we believe that individual differences are likely to be magnified and accentuated during periods of discontinuity, as each individual, in an effort to regain control over the changing situation, actively attempts to assimilate discrepant events into existing cognitive and action structures. Periods of social discontinuity thus provide a unique opportunity for discerning the principles that govern behavioural continuity.

The challenge of change

We have suggested—perversely—that potentially disruptive social transitions in the life course produce continuity, not change. When, then, does change take place? Have we eliminated, through theoretical fiat, the most promising candidate?

No. We have instead gained a finer appreciation of the very special conditions under which *systematic* change may be promoted and produced. To effect change, new situations have to be all-encompassing, pressing the individual 'from every angle toward a particular type of development or outcome' (Anthony 1987, p 35). To invoke the earlier distinction borrowed from experimental research, weak situations must be converted into strong situations in order to promote change and growth in human development. Instead of permitting to flourish the type of ambiguity that characterizes our control groups, we shall have to deliver the type of manipulations that we typically use with our experimental groups.

Indeed, one reason it has proved so difficult to replicate in real-world settings the type of malleability that is so easily demonstrated in laboratory settings is that real people seek out opportunities that promote continuity and consistency. Laboratory settings are effective because they limit the ways in which subjects can structure the environment in which they behave. In an ingenious series of studies, Swann (1983) has thus shown that people do change, but only when they receive self-discrepant feedback in highly structured situations where they have little opportunity to resist the treatment they receive. Even this, however, may not be enough. Consider efforts to suppress undesired behaviour (Moffitt 1983). Research on the learning theory model of punishment suggests that punishment is most effective in suppressing behaviour when alternative behaviour is made available (Azrin & Holz 1966). Unless old options are challenged or eliminated, and unless new opportunities are created and new scripts provided, individuals will often resort to familiar techniques of self-verification rather than engage in the more arduous efforts aimed at change.

These data support Bloom's (1964, p 212) contention that 'it is the extent to which a particular solution is overdetermined that makes for a powerful environment' capable of inducing meaningful personal change. They also converge with evidence from recent life-course analyses of turning points in behavioural development.

Consider the case of deviant youth. What distinguishes the natural histories of men who are 'life-course persistent' from those who defy the probabilities of a cumulative sequence of life disadvantage? Sampson & Laub (1990) addressed this question in a longitudinal study of 500 delinquents and 500 non-delinquent controls. The answer points to a set of informal social controls that serve to reduce involvements in familiar and deviant social habits and that simultaneously assign new behavioural scripts. In particular, job stability and marriage both modify trajectories of deviance across the life course. Importantly, these results cannot be attributed to self-selection because they were obtained in analyses performed separately among delinquents and non-delinquents and which controlled for original level of deviance.

Consider also the case of children who grew up during the Great Depression. Elder (Elder 1979, Elder & Caspi 1990) has examined the historical context of two birth cohorts who lived through the Great Depression, World War II, and the Korean campaign: the Oakland Growth sample of 167 members (birth dates, 1920–1921) and the Berkeley Guidance sample of 214 members (birth dates, 1928–1929). Both studies have been carried out at the Institute of Human Development, University of California, Berkeley. Data were collected on these cohorts across the 1930s on an annual basis and during widely spaced follow-ups up to the 1980s.

A comparison of these samples offers a rare opportunity to examine age and sex variations in vulnerability to stress: namely, family income loss during the Great Depression. Members of the Oakland study were beyond the critical years of dependency during the Great Depression and they left school during a period of rising prosperity. In contrast, members of the Berkeley study were less than two years old when the economy collapsed and they remained exclusively within the family through the worst years of that decade.

The effects of economic deprivation on children were quite variable. Oakland adolescents were often called upon to assume major family responsibilities when the economy collapsed. Boys, in particular, sought work outside the home, whereas girls assumed major household tasks within the family. Family change of this sort enhanced the independence of boys. For girls, however, greater family involvement meant greater exposure to tension and, not surprisingly, was associated with some adjustment difficulties in adolescence. Among the Berkeley children, these sex differences reverse. Among girls, the short-term effects of deprivation were negligible. However, the findings show that economic deprivation and family stress was most pathogenic for boys in early childhood.

How did these boys withstand and rise above their misfortune? Elder (1986, Elder & Caspi 1990) suggests that part of the answer centres on military experience as a turning point. Men who grew up in deprived circumstances could break the cycle of disadvantage by entering the service, especially by doing so at a relatively young age.

Three features of the military experience appeared to promote *psychological* change. The first is the 'knifing off' of past experience. Induction meant separation from the immediate influence of family and community; it made prior identities irrelevant. A second feature is the extent to which military service represented a psychosocial moratorium. Military duty provided a legitimate 'time out'; it released the recruit from conventional expectations and from the commitment pressures that Erikson describes in the concept of role diffusion. A third feature is the broadened range of social comparisons that provided an opportunity for personal development through the setting of new achievement goals.

This is not an exhaustive list of the features of the transition to military life, but together they define a developmental pathway that may have offered a promising route to life change. Together they also highlight the role of strong situations in reducing pre-transition differences between individuals and in restructuring post-transition opportunities for them.

Conclusion

Recent longitudinal research has taught us to treat the stability and consistency of individual differences as a problematic phenomenon requiring both confirmation and explanation. No longer can we assume implicitly that continuities across time and circumstances are simply manifestations of intrapersonal dispositions. We have learned instead to seek the explanation for both continuity *and* change in the interaction between the person and the environment.

This search has revealed some surprises. Continuity is not guided by inertia; rather, there are vigorous processes that promote continuity across the life course. Moreover, these processes appear to be enhanced in settings that are characterized by novelty, ambiguity, and uncertainty. Contrary to our assumptions, behavioural changes are not direct outcomes of environmental changes. In fact, behavioural *continuity* is more likely to emerge during periods of social discontinuity.

Change is not impossible, of course, but we are learning that the situational requisites which define turning points in behavioural development are more rigid and specialized than was previously thought. The processes promoting continuity are aggressive foes. And those of us in the mental health professions must now recognize that in order to promote salutary outcomes we shall have to do more than provide opportunities for change. We also have to eliminate those opportunities that allow active processes of continuity to flourish and to guide the trajectory of the life course.

Acknowledgement

This work was supported in part by a grant from the National Institute of Mental Health (MH-41827).

References

Anderson CR 1977 Locus of control, coping behaviors and performance in a stress setting: a longitudinal study. J Appl Psychol 62:446–451

Anthony EJ 1987 Risk, vulnerability, and resilience: an overview. In: Anthony EJ, Cohler BJ (eds) The invulnerable child. Guilford Press, New York, p 3–48

Azrin NH, Holz WC 1966 Punishment. In: Hongi WK (ed) Operant behavior. Appleton-Century-Crofts, New York, p 60–79

Bloom BS 1964 Stability and change in human characteristics. Wiley, New York

Bowlby J 1973 Attachment and loss, vol 2. Basic Books, New York

Buss DM 1985 Human mate selection. Am Sci 73:47–51

Buss DM 1987 Selection, evocation and manipulation. J Pers Soc Psychol 53:1214–1221

Cairns RB, Cairns BD, Neckerman HJ, Gest SD, Gariepy J-L 1988 Social networks and aggressive behavior: peer support or peer rejection? Dev Psychol 24:815–823

Cantor N, Kihlstrom J 1987 Personality and social intelligence. Prentice-Hall, Englewood Cliffs, NJ

Caspi A, Bem DJ 1990 Personality continuity and change across the life course. In: Pervin L (ed) Handbook of personality theory and research. Guilford Press, New York, p 549–575

Caspi A, Herbener ES 1990 Continuity and change: assortative marriage and the consistency of personality in adulthood. J Pers Soc Psychol 58:250–258

Caspi A, Moffitt TE 1991 Individual differences and personal transitions: the sample case of girls at puberty. J Pers Soc Psychol, in press

Caspi A, Bem DJ, Elder GH Jr 1989 Continuities and consequences of interactional styles across the life course. J Pers Soc Psychol 57:375–406

Darley J, Fazio RH 1980 Expectancy confirmation processes arising in the social interaction sequence. Am Psychol 35:867–881

Dodge KA 1986 A social information processing model of social competence in children. In: Perlmutter M (ed) Minnesota symposia on child psychology. Erlbaum, Hillsdale, NJ, vol 18:77–125

Elder GH Jr 1979 Historical change in life patterns and personality. In: Baltes PB, Brim OG Jr (eds) Lifespan development and behavior. Academic Press, New York, vol 2:117–159

Elder GH Jr 1986 Military times and turning points in men's lives. Dev Psychol 22:233–245

Elder GH Jr, Caspi A 1988 Economic stress: developmental perspectives. J Soc Issues 44:25–45

Elder GH Jr, Caspi A 1990 Studying lives in a changing society: sociological and personological explorations. In: Rabin AI, Zucker RA, Emmons RA, Frank S (eds) Studying persons and lives. Springer, New York, p 201–247

Entwisle DR, Alexander KL 1988 Early schooling as a 'critical period' phenomenon. In: Namboodiri K, Corwin RG (eds) Sociology of education and socialization. JAI Press, Greenwich, CT, vol 8:27–55

Epstein S 1980 The self-concept: a review and the proposal of an integrated theory of personality. In: Staub E (ed) Personality: basic issues and current research. Prentice Hall, Englewood Cliffs, NJ, p 81–130

Feldman KA, Newcomb TM 1969 The impact of college on students. Jossey-Bass, San Francisco

Fuller JL 1967 Experiential deprivation and later behavior. Science (Wash DC) 158:1645–1652

Greenwald AG 1980 The totalitarian ego: fabrication and revision of personal history. Am Psychol 35:603–618

Ickes W 1982 A basic paradigm for the study of personality, roles and social behavior. In: Ickes W, Knowles E (eds) Personality, roles and social behavior. Springer, New York, p 305–341

Markus H 1977 Self-schemata and processing information about the self. J Pers Soc Psychol 35:63–78

Matheny AP Jr, Dolan AB 1975 Persons, situations, and time: a genetic view of behavioral change in children. J Pers Soc Psychol 32:1106–1110

Mischel W 1977 The interaction of person and situation. In: Magnusson ID, Endler NS (eds) Personality at the crossroads: current issues in interactional psychology. Erlbaum, Hillsdale, NJ, p 333–352

Moffitt TE 1983 The learning theory model of punishment: implications for delinquency deterrence. Crim Justice Behav 10:131–158

Monson TC, Hesley JW, Chernick L 1982 Specifying when personality traits can and cannot predict behavior: an alternative to abandoning the attempt to predict single-act criteria. Pers Soc Psychol 43:385–399

Patterson GR 1982 Coercive family process. Castalia, Eugene, OR

Patterson GR, Bank L 1990 Some amplifying mechanisms for pathologic processes in families. In: Gunnar M (ed) Minnesota symposia on child psychology. Erlbaum, Hillsdale, NJ, vol 22:167–209

Plomin R, Rowe DC 1979 Genetic and environmental etiology of social behavior in infancy. Dev Psychol 15:62–72

Plomin R, DeFries JC, Loehlin JC 1977 Genotype–environment interaction and correlation in the analysis of human behavior. Psychol Bull 88:245–258

Rutter M 1983 Statistical and personal interactions: facets and perspectives. In: Magnusson D, Allen VL (eds) Human development: an interactional perspective. Academic Press, New York, p 295–320

Sampson RJ, Laub JH 1990 Stability and change in crime and deviance over the life course: The salience of informal social ties. Am Soc Rev 55:609–627

Scarr S, McCartney K 1983 How people make their own environments: a theory of genotype–environment correlations. Child Dev 54:424–435

Snyder M 1984 When beliefs create reality. In: Berkowitz (ed) Advances in experimental social psychology. Academic Press, Orlando FL, vol 18:248–305

Snyder M, Ickes W 1985 Personality and social behavior. In: Aronson E, Lindzey G (eds), Handbook of social psychology. Random House, New York, vol 2:883–947

Swann WB Jr 1983 Self-verification: bringing social reality into harmony with the self. In: Suls J, Greenwald AG (eds) Psychological perspectives on the self. Erlbaum, Hillsdale, NJ, vol 2:33–66

Thiessen D, Gregg B 1980 Human assortative mating and genetic equilibrium: an evolutionary perspective. Ethol Sociobiol 1:111–140

Thomas WI, Thomas D 1928 The child in America. Knopf, New York

Tomkins SS 1986 Script theory. In: Aronoff J, Rabin AI, Zucker RA (eds) The emergence of personality. Springer, New York, p 147–216

Wachtel PL 1977 Psychoanalysis and behavior therapy. Basic Books, New York

Wilson EO 1975 Sociobiology. Harvard University Press, Cambridge, MA

Wright JC, Mischel W 1987 A conditional approach to dispositional constructs: the local predictability of social behavior. J Pers Soc Psychol 53:1159–1177

DISCUSSION

Barker: It is fascinating how certain themes run through each paper—ideas about critical periods at which a good or bad environment can have its maximal effect, and ideas that adaptations to an adverse environment may not only be long term but may also be beneficial. (It is also interesting to hear data on a normal human population rather than on an extreme population, like, for example, the premature babies.)

Dr Caspi, it would be interesting to know the long-term effects on the physical health of members of your groups of men and women from deprived backgrounds.

Caspi: The answer to this question may prove somewhat surprising. Some of our preliminary analyses, based on self-reports of health status at midlife, reveal that health problems were most often reported by successful men who grew up in deprived circumstances; that is, by men who managed to surmount a childhood of disadvantage.

Suomi: That result doesn't sound like much of a success story! One needs to know, however, the nature of the self-reports, and how many of those cases might actually translate into clinical cases.

Caspi: Self-reports of physical health are prone to many problems. The participants in this study were asked to rate whether they were suffering from any health difficulties, to evaluate whether they had any drinking problems, and to indicate, in a general way, their level of fatigue. If our preliminary results hold up under closer scrutiny, there are two possible explanations for the excess of health problems among men who have done rather well in their lives, given their child-hood socialization in deprived circumstances. The first explanation draws on the sociological hypothesis of 'family dissociation'. According to this hypothesis, one consequence of upward mobility is a strong ambivalence toward one's origins. This ambivalence may be difficult to face at a conscious level; it may, then, be turned inward and 'somatized'. A second explanation, and a more reasonable one, focuses on the unique life histories of successful men who managed to surmount a childhood history of disadvantage. Unlike their equally successful but non-deprived counterparts, their road to success was not altogether straightforward. In this group were found men who did not finish high school, who went into the army at a young age, and who then resumed and completed their education years later. Their life careers were characterized by many uncertainties in young adulthood. In short, their path to success was protracted and circuitous. Rather than being muted by rewarding achievements, it is possible that health problems were actually an outcome of expending so much energy in early adulthood attempting to overcome many obstacles.

Suomi: The military service that you discussed was presumably associated with participation in one of the so-called 'good' wars (such as World War II). My generation, reared in the 1950s, was also exposed to military service, not entirely voluntarily in most cases, and, from all the anecdotal information I can get, the usual outcome was not so positive. One has here the basis for a nice model, at comparable points in life, in comparable (military) institutions, but with very different outcomes. It would be interesting to do a survey starting in 1950, rather than 1920 or 1930, and to see how that would affect your conclusions. What I am suggesting is the possibility that entry into new groups may also have largely *deleterious* effects, because of the nature of the groups or situations, rather than the *positive* effects of entry into a new experience— military service, in your studies—that you have suggested.

Caspi: You have raised two points. Your point about historical or cohort differences is very important. We simply don't have very good comparative data. Your point about 'deleterious effects' of entry into new groups warrants a brief elaboration. I agree that many new experiences can have deleterious effects. Novelty, ambiguity and uncertainty are, in my view, situational conditions that serve as a catalyst for behavioural consistency. In fact, we've argued that these situational conditions serve to accentuate old, familiar ways of behaving. The way, then, to produce behavioural change is to reduce the degree of ambiguity in a novel situation. Thus, if you create a 'press' to behave, but fail to provide a behavioural 'script' for *how* to behave, you will see accentuation and retrenchment of familiar behaviour patterns, much as you do in species displacement. What one wants to do is to provide the individual with a new behavioural 'script' and to 'disambiguate' the novel situation. My point is that a total institution like the military does precisely this.

Lucas: You suggest an intrinsically psychological explanation for higher morbidity in middle age in some of the groups from a deprived background, but the fact is that you have seen an apparent increase in morbidity in children from a deprived background. The question is the extent to which that is due to reporting bias or due to a genuine increase in pathology. That is central to what we have been talking about from David Barker's work, for example.

Caspi: The self-reports of health difficulties are confined to a very specific group: to those men who managed to *surmount* early adversity. Indeed, these health difficulties did not characterize those men who grew up in deprived circumstances and who did not do particularly well in terms of social achievement in adulthood. Let me caution, however, that these particular results are based on very small subsamples. To amplify Dr Rutter's insights, these results are instructive in suggesting that long-term outcomes associated with different childhood environments are not simply either adverse or salutary; they are often a bit of both. As I noted, in my work with Glen Elder we've seen that men who made great strides in their social attainment and psychological health despite a childhood of deprivation suffered some costs in terms of physical health.

Barker: The question is whether they are costs or whether they were determined by what happened in childhood.

Golding: If you had information on morbidity before these men went to college, you might see the same pattern then, which would be a *continuity* in morbidity that had nothing to do with going to college (except in as much as children who are less mobile during childhood may be more likely to divert their energies into reading, and hence will tend to go to college).

Caspi: This is a very good point. There is a possibility that the health costs of growing up in deprived circumstances were already present earlier in childhood and adolescence and then re-emerged again later in adulthood. My impression from having examined the health data on the longitudinally studied participants during adolescence is that there was no significant association between childhood deprivation and physical health during the second decade of life. In addition, this 'continuity of morbidity' thesis, although very provocative, does not account for the increase in physical problems among men who *overcame* the difficulties associated with a deprived childhood.

Wadsworth: We have found a 'cost' of good educational attainment, in the sense that women who moved from low socio-economic circumstances in childhood to high occupational achievement, because of their educational attainment, had a significantly higher average consumption of alcohol and fats compared with other women (Braddon et al 1988).

Murray: Dr Caspi, I wasn't clear whether you think that the groups in your cohort studies differed as a consequence of having gone to college, or whether they reached college as a consequence of already being different.

Caspi: This is a very critical distinction and it's difficult to tease apart in our data. Although it is hard to establish a causal sequence with the relatively crude temporal categories at our disposal, let me try to patch together the most plausible interpretation of the longitudinal findings that I've had the privilege of analysing with Glen Elder.

Men who grew up in deprived circumstances were significantly more likely to join the military at a relatively young age. We also know from personality assessments in adolescence that these men were significantly less goal-oriented, less socially competent, and felt more inadequate than men who did not join the service or did join it but at a later age. Now, the military served two purposes: first, we know that, from adolescence to adulthood, men who entered the service early experienced the most profound psychological change; in particular, they became much more poised and competent. Second, whatever personal changes may have occurred during military service, these changes were accompanied by the opening-up of new opportunities. Indeed, the social impact of military service among deprived men was, in part, indirect; it not only produced psychological change, it also opened up educational opportunities for them.

The answer to your question is thus somewhat complicated. There was a selection effect into college, because there was a selection effect into military service at a relatively young age. But the selection effect into military service is not at all 'intuitive'; it involved incompetent boys heading into a new environment that could perhaps—and often did—remove them from a disadvantaged life.

Lloyd: You haven't said what the *women* in the birth cohorts did in lieu of going into the army. What changes did they show?

Caspi: As I mentioned in my paper, the adverse effects of economic deprivation on girls were found primarily among those who were adolescents during hard times. How did these girls fare in later life? Physical attractiveness is a key factor to understanding the variability in outcomes among these women. In an analysis of members of the Oakland Growth Study, we showed that to understand the effects of economic hardship on girls' lives requires that we attend to the 'buffering factor' of physical attractiveness (Elder et al 1985). In particular, we learned that attractive girls (as judged by physical features rated by independent observers) were not likely to be ill-treated by their fathers, no matter how severe the economic pressure. This finding underscores the importance of viewing the developmental effects of environmental change in relation to the child's characteristics. In addition, we know that these physically attractive girls were significantly more likely to marry more successful men, a finding that is in keeping with much of what we know about hypergamous marriages in the Western world; physical attractiveness is one of the best predictors of upward mobility through marriage among women. It's also a mating preference that characterizes men in societies throughout the world (Buss 1989). But I think our results imply something more than just physical attractiveness *per se*; our results indirectly highlight the importance of relationships in women's lives. And physical attractiveness—for better or worse—contributes to the quality of relationships women have with men, be they fathers or potential mates.

Richards: There are, of course, cultures where men and women don't see each other before they are married!

May I ask about the men's marriages in the cohorts? Could it be that marriage mediates some of the effects of the army and of education on outcomes? The army and education could have quite profound effects on the choice of partner and age of marriage.

Caspi: This is a very exciting possibility that I think we shall have to pursue more carefully. Unfortunately, I can't really answer that with the available evidence because we don't have a very clear-cut measure of the kinds of resources that the wives offered men. In terms of measurement, it is harder to identify a supportive spouse than it is a rich one.

Richards: Was the age at marriage different in the various groups?

Caspi: One would expect those men who grew up deprived and ended up doing well in their lives to have married at a later age.

Rutter: That is surely what your data showed, that one of the effects of military service was to postpone marriage?

Caspi: You're right. We know that men who entered military service at an early age were more likely to postpone their marriages to an older age. What I don't know is whether this marital delay also accounts for the relationship between early military service and occupational success among men who grew up deprived. This is a very interesting possibility.

Richards: I would also expect the men who do well to 'marry up', socially; they may be more likely to find a female partner matched to their own educational attainment, so they have more highly educated wives than those who married earlier. You emphasize that women get somewhere because of marriage, but that is a one-sided view; I think men's careers also have quite a lot to do with who they choose to marry.

Caspi: No doubt, somewhere along the way new relationships did promote further change.

References

Buss DM 1989 Sex differences in human mate preferences: evolutionary hypotheses tested in 37 cultures. Behav Brain Sci 12:1–49

Braddon FEM, Wadsworth MEJ, Davies JMC, Cripps HA 1988 Social and regional differences in food and alcohol consumption and their measurement in a national birth cohort. J Epidemiol Community Health 17:525–529

Elder GH Jr, Van Nguyen T, Caspi A 1985 Linking family hardship to children's lives. Child Dev 56:361–375

General discussion

Barker: There seems a mass of evidence which suggests that adaptation to the environment in fetal life and early childhood has major effects on the risk of developing a number of important diseases. That this is the case in psychological disease has long been known. It seems likely to hold for cardiovascular disease, where early findings from geographical studies have been strikingly borne out in individuals. There are now strong links between early adaptations and lifetime patterns of allergic response. The evidence for the infant origins of chronic obstructive lung disease is increasing. And there are new findings pointing to the fetal and childhood origins of schizophrenia and motor neuron disease. No doubt there are other diseases in which early life has not been examined and for which strong relationships will be found.

What are the processes by which early responses to the environment bring about such apparently large changes in adult health and life expectancy? We know there are circumstances where the early environment has massive effects, such as in chronic rheumatic heart disease. We seem to know quite a lot about the links between psychosocial adversity and altered responses to social challenges. We know that certain childhood infections have major effects on adult diseases in the airways, the liver and possibly the CNS. In the past three days we have heard that early adaptations affect such various processes as the connections between nerve cells, the setting of blood pressure, the motor development of premature babies, the hepatic metabolism of cholesterol throughout life, and immune responses. The question is where we go on from here. Clearly we need more research into possible links between the environment and later disorders. This is an area where funding is notably difficult to obtain. We also need to know more about the biological processes underlying known links: without such knowledge we shall be unable to make effective interventions.

Are there particular areas where it is now clear that a lot more research should be done, in the interests of the public health?

Hamosh: Let me take up that point. Now, for the first time in the US, the government is interested in the effect of nutrition on pregnancy and lactation, and has requested the National Academy of Sciences to prepare reports on this topic (National Academy of Sciences 1990, 1991). There is a good deal of information on nutrition during pregnancy and its effect on the outcome of pregnancy. There is, however, very little on the effect of nutrition during lactation, especially in developed countries. It is extremely important to know what those effects are.

We know that milk consumption by the newborn is 'infant driven'; in other words, maternal milk production is regulated by the infant, and not by the

224

mother. Very convincing studies of milk consumption have been done in three different areas of the US, one in Colorado (Neville et al 1988), where there might be an effect of high altitude; one study in Texas which has an American population and also a Hispanic population of Mexican origin (Butte et al 1984); and a third study of an affluent upper middle-class group in California (Dewey et al 1991). From these studies we know that the infant regulates the amount of milk that its mother is producing and secreting, by means of a very strong feedback regulation on what the mother will synthesize and secrete (Dewey & Lonnerdal 1986).

We also know that size at birth defines the amount of milk consumed (Prentice et al 1986); therefore babies who are big at birth and are well-developed have the best chance of favourable postnatal growth. We come back to the question of how to influence the fetus to grow to the right size and afterwards to receive the best possible nutrition postnatally. There might be several factors that affect this aspect; maternal nutrition might be one of them. The possible effects of neurotransmitters on brain development, the role of the satiety centre, etc. are all things to be taken into consideration.

We are also looking in the US at different ethnic groups. We have two large immigrant groups that are relatively new to the country; a very strong influx of Hispanics from Mexico and other central American countries, and a considerable influx of immigrants from Asia. These groups have different dietary habits from the general population. It is interesting that although one tries to adapt to the new country in one's diet, when one is pregnant one reverts to one's original habits! I don't know whether this is the influence of one's mother, or grandmother, or some continuity of tradition. The importance of this behaviour for us in the US is to see what the dietary habits are and what foods might specifically be missing and need to be supplemented. Is the food intake in accord with recommended dietary allowances (National Academy of Sciences 1989) for pregnant and for lactating women in the US? Those are new problems for us, and it is encouraging that we are now looking at these questions.

The problem of funding such studies arises. These need to be large cohort studies, which have to be longitudinal, because one has to look at infant outcome; one also needs to look at maternal outcome, because we know that during lactation, the concentration of certain nutrients decreases in milk; selenium is such an example (Smith et al 1982). Zinc is another nutrient that decreases in the course of lactation (Casey et al 1989). One can increase milk selenium levels by supplementing maternal nutrition with selenium (Mannan & Picciano 1987). However, zinc supplementation of the maternal diet does not raise zinc levels in milk (Moser & Reynolds 1983). There is also a suspicion that zinc might determine the appetite of the infant. The fact that breast-fed infants tend to drop off from the normal growth curves after four months or so of exclusive breast feeding indicates that their dietary intake is lower than that of formula-fed infants. It could be that zinc becomes limiting at this age

(Neville & Oliva-Rasbach 1987). Zinc is a component of many enzymes and regulatory proteins, so this may be important, but we don't know yet. There is need for much research in this vital area of infant development.

Barker: You are advocating research not just on what women need when lactating or during pregnancy, but before that?

Hamosh: Yes, much earlier. The crucial period to examine is women's nutritional habits before they become pregnant. We also have a large population of pregnant teenagers in the US, starting at the age of 10 or 11! These young girls often do not even know they are pregnant until delivery. They are quite uninformed and unaware of the nutritional needs of pregnancy. The question is: what are the effects on the fetus when its mother is still growing? She is herself a child, and perhaps there are completely different nutritional requirements at this time. This is a growing segment of the US population. We need to pay attention to what girls eat early in life, starting with health education in the elementary school, perhaps giving questionnaires once or twice a year to get general information about what young girls are eating.

Barker: Is it fair to say that enough is known about short-term harmful outcomes in relation to poor maternal nutrition, and poor nutrition in girls, to make this an important area for health education? And if, as some of us suspect, the effects of maternal nutrition and physique are immensely important in the long-term in relation to cardiovascular and other diseases, does this simply heighten the problem? There is clearly a lack of knowledge in many areas, but is enough known to make immediate new recommendations?

Hamosh: Enough is known for general recommendations, but we must go beyond this stage and discover what we can do and what policies we need to adopt, to improve the nutrition of malnourished girls and women, preferably before they get pregnant.

Murray: Perhaps as well as thinking what mothers don't get, nutritionally, before or during pregnancy, we should think what mothers *are* now getting, in view of the subtle process of dendritic organization in the brain. We need to know what happens when a mother takes LSD or heroin or other psychoactive drugs during pregnancy. Perhaps we should be alarmed about the frequency with which teenage mothers are taking concoctions of drugs that may interfere with fetal brain development.

Blakemore: Rather little is known about the impact of drugs taken during fetal life (apart from alcohol and tobacco) on the later development of the offspring, but this raises an interesting question. At this meeting we have been concerned largely with specific sensitive periods and with the permanent impact of events early in life on later health. If we could discover what conditions make a system sensitive at a particular stage in development and not later on, we might eventually be able to replicate, at any time in life, the conditions that lead to sensitivity. This might be dangerous; but it might be beneficial. As far as the influence of experience on the visual cortex is concerned, there are strong hints that sensitivity early in life depends on substances (perhaps including acetylcholine, noradrenaline, growth factors, etc.) acting on cortical neurons through specific membrane receptor systems (especially the NMDA receptor) which are selectively expressed at an early age. Increased knowledge of the factors controlling early growth and sensitivity might eventually give us the possibility of restoring those conditions and of remedying the bad consequences of early vulnerability.

Barker: You stressed in your paper that, given a certain genetic constitution, the organism is highly sensitive to its environment and moulded by it. This is a process with considerable evolutionary advantage, and a basic property of many organisms and systems.

Blakemore: Yes, such processes are simply more exaggerated in the mammalian brain. Clearly, the growth and differentiation of many parts of the body are influenced by the environment, including the internal environment. The environment provides many of the factors that determine the differentiation of tissues and also the differentiation of individuals.

Casaer: Although most neurons in the adult mammalian brain are postmitotic and thus not capable of dividing, Nottebohm demonstrated that neurons involved with the songs of birds in their adult sexual life can increase in number. This finding revolutionized our thinking on plasticity and adaptive possibilities in adult neurology and its probably important neuroendocrine control (Paton & Nottebohm 1984).

Dobbing: I am anxious that we do not give an oversimplified picture, particularly of nutrition. For example, as regards lactation, it's generally true, at least for macronutrients, that the influence of gross variations in maternal nutrition is not on the quality of the breast milk, but on its quantity. You can go into refugee camps, or to the Third World, and find marasmic babies at the breast; analysis shows that the quality of that milk is largely the same as in the well-fed mother. What has changed is the quantity. It would be better, even if it were not absolutely true in every detail, to think of this from a less complicated position: that what matters is the quantity of breast milk, its quality having been decided (I am excepting nutrients like zinc, of course). This matters because it largely determines the gross growth rates of the infant, particularly in terms of length as an index of growth, and similarly through into childhood. It is much less useful, and almost certainly false, to say that unless pregnant and lactating women eat a particular fat, say, they are putting their children's intelligence in jeopardy. This type of statement is misleading and I would make a plea for responsibility on our part here. Such statements eventually find their way to the uninstructed public and cause unnecessary anxiety, and even guilt.

Hamosh: I agree that what is important is adequate nutrition. Fortunately, the process of lactogenesis is very robust and is continuing in women who are severely deprived of food. The volume of milk may change; milk quality might also change; the concentration of micronutrients, macronutrients and bioactive components may change too. For example I would not want to feed my own child milk that contains, let's say, 80% of medium-chain fatty acids synthesized from carbohydrate because in such large amounts they are not the best food for the infant. You cannot use medium-chain fatty acids to synthesize cell membranes. So there are certain drawbacks.

Dobbing: Do you say this from the results of studying actual outcome, such as behavioural or growth studies, or merely theoretically? I think measurable actual outcome is much more interesting and reliable than the theory of what 'ought' to happen!

Hamosh: Thorough human outcome studies are much needed. There are animal and human studies which show that one would like to have a balanced diet, theoretically at least (see e.g. this volume: Lucas 1991). The mammary gland can synthesize some nutrients and the mother can thereby to a certain extent compensate for nutritional inadequacies. Fortunately there is this plasticity in the system. We have, however, paid no attention to the mother's condition after nursing her child, when she is not sufficiently fed; how does this affect her body reserves, in relation to her next pregnancy and lactation? What are the long-term effects on her health, or bone status? We have to know much more before we can make proper recommendations.

For example, many people have the misconception that human milk is high in cholesterol. This is based on old, incorrect analyses that indicated that human milk contains about 100 mg per dl cholesterol, when it in fact contains one-tenth of that amount, according to newer analytical procedures (Bitman et al 1986, Jensen 1989).

Lucas: My view is that we are not at the stage, for infant nutrition, of advocating what is best. If we are looking for models of physiological nutritional behaviour, we have lost sight of Nature completely here, because we have so contaminated infant feeding practices with Western advice that we no longer know what is 'normal'. In any case, we may not now be interested in what once evolved, because our biological targets in modern society have changed and we now care about longevity, freedom from disease in old age, and postreproductive survival—all of which might have been quite unimportant when breast feeding was evolving. What we need are data on the outcomes of modern practice. The lack of such data in fields like nutrition (and, I suspect, many other fields) has resulted in the continually changing patterns of advice given to parents. In this century, every five years or so there has been a major change in views, so that the public health professions are utterly confused over what they should be doing. This is because we have lacked outcome data, because we have not made a commitment to answering the questions that people want answered. We have become preoccupied by the short-term physiology and by the exploration of mechanism, which we find intellectually exciting; but we haven't answered the practical questions. We should address those first, define the important events and then look for mechanisms to explain them. I therefore agree very much with what John Dobbing is saying.

Barker: You are saying we need to know long-term outcomes of maternal and infant nutrition, but you are not suggesting that we wait on those before addressing the short-term physiology? Surely the two must go on in parallel.

Lucas: As a physiologist, I am of course interested in learning what the physiological mechanisms are, but the order of priority now seems to be to look at nutritional outcomes, since this has been so neglected, and then try to understand the mechanisms involved. There is no point in having a wonderful theory about an outcome event that may not even occur.

Barker: You use preterm babies as a model. To the outsider, the problem is that not only are you studying fetuses which have to make their way through all the uncertainties that all fetuses face, with the problems of being born at a particular point in gestation, which may advantage the lungs but disadvantage the brain, or vice versa, but you are experimentally imposing a lot more on your infants. For example, you might be shortening timescales which should be longer by giving the preterm babies certain kinds of food, and myelinating their axons earlier.

Lucas: Clearly, with babies in an unphysiological situation, where critical signals are being changed by environmental circumstances, it is possible that we (as physicians, or experimenters, or both) may be programming these preterm infants in a way that would not have occurred had they been born at term. But that may not be so; we might be looking at an exaggeration of what would have occurred at full term. Nevertheless, with preterm babies, still effectively fetuses, in the outside world, one must be changing normal developmental sequences, and many events will happen too soon. Gluconeogenesis is switched on months before the normal time; the gut is also doing things that it would not normally be doing at that time—for instance, lactase activity is suddenly switched on at 28 weeks rather than 40 weeks. Whole systems are operating in an unusual way; the cardiovascular system moves into postnatal mode from fetal mode months too soon. So the premature baby is perhaps a good model for investigating what happens when switches are thrown at the wrong time.

Barker: Are there any benefits to prematurity? It's perhaps not ideal for every system that 40 weeks is the period of human gestation.

Lucas: We haven't identified them, although we haven't studied our population of premature babies long enough.

Rutter: A common thread in all presentations has been the necessary attempt to infer underlying mechanisms from statistical associations. As Alan Lucas and David Barker have emphasized, this process of inference usually requires a combination of experimental and epidemiological data. Risk indicators have to be differentiated from risk mechanisms. A striking example in the psychosocial arena is provided by Brown & Wing's (1962) study of mental hospitals, in which it was found that one of the best measures of an effective institution was whether or not patients had their own toothbrush and some article of personal ornament, such as a ring or a watch! Of course, in this instance, it was obvious that this was only an indicator; no one would suppose that handing out toothbrushes and ornaments in a bad institution would improve outcomes! However, sometimes it is not so easy to tell. Moreover, in the case of many psychosocial risk factors, laboratory experiments are not feasible. In these circumstances, experiments of Nature may still provide a good epidemiological alternative (Rutter 1981). The challenge for the epidemiologist is to seek situations where contrasting patterns enable variables to be 'pulled apart' in ways that approximate to the contrived laboratory experiment.

A further issue that has come up in several sessions concerns the relative strength of different causal variables. In that connection it is necessary to remind ourselves that there are several quite different ways in which the causal question can be proved (Rutter 1979/80). Much research focuses on the cause of individual differences—why John has a particular disorder and Bill does not. That is an important issue but it may be equally important to examine the causes of changes in prevalence over time; the answer may not be the same. Thus, personal factors are important in accounting for individual differences in unemployment but they played no part in the massive rise in unemployment that occurred in the UK between the 1970s and mid 1980s. Jack Tizard (1975) drew attention to the observation that the average height of London schoolboys rose by about 10 cm during the first half of this century (almost certainly as a result of improvements in nutrition) but this effect did not alter the predominance of genetic influences accounting for individual variations in height at any one time. We need to be alert to the varied nature of causal questions and their answers.

In considering individual differences, I have been struck throughout this symposium by the pervasiveness and importance of person–environment interactions. We have had several accounts of effects in one population subgroup but not others. A nice example came from Alan Lucas' examination of the effects of diet on atopy, where the overall findings revealed no effect, yet there was an effect in those with a positive family history. There are important implications for styles of data analysis (Rutter & Pickles 1991).

In this session we have argued about the implications for policy and practice of our research findings. There are twin dangers of excessive caution and of premature action based on a faulty understanding of the processes involved. We are all aware of scientists who have made fools of themselves by advocating sweeping policy changes that were seen as foolish when better evidence became available. Their experience serves as a warning that must be heeded but, equally, there is a danger that we do nothing until we know everything—a point that never comes. The evidence that has been presented here, in both the somatic and psychosocial arenas, does provide the basis for taking at least some cautious steps forward on policy and practice.

Wadsworth: There are many long-term follow-up studies containing much information on what we take to be environmental influences. Knowledge accruing in the human genome mapping projects now offers huge possibilities for longitudinal studies to compare the influence of environment and of genetic factors. That is a resource that we shall be able to make much more use of in the future.

Hanson: Incidentally, pregnancy is one time when the maternal rather than the paternal genome is extremely important, in terms of growth. Walton & Hammond (1938) showed that when Shire horses were crossed with Shetland ponies, the size of the offspring at birth depended on the maternal genome. One musn't forget the interaction between genetic elements and nutrition during pregnancy.

Also, David Barker's correlations make us ask about the interaction between fetal size, or fetal growth rate, and placental size or rate of growth. We know virtually nothing about what controls the growth of the placenta and whether it is under fetal or maternal control or a combination of the two; is it nutritional, is it genetic? We have to be cautious, simply from a physiological point of view, about pointing the finger solely at nutrition during pregnancy here.

Chandra: The genetically determined contribution to stature is equally contributed by the mother and father, surely?

Hanson: Yes, eventually that is true, when postnatal growth has occurred; but I'm referring to prenatal growth. Size at birth has been reported to be more dependent on the maternal than the paternal genome, and it might have important consequences, as is clear from David Barker's data.

Mott: It is important to be as quantitative as possible in our assessments, because in our society there are immense distortions of relative risk; people exaggerating some risks (such as traces of pesticides in food) and ignoring other factors that are of much more consequence (such as alcohol abuse). In the US alone, alcohol abuse causes about 25 000 deaths on the highways and a loss in wages and health care costs of billions of dollars per year. This is just one example of where we have completely distorted the relative risks and costs. As scientists we tend to do the same, so I would emphasize the need to achieve a quantitative description of the risk or benefit of any kind of intervention.

Barker: In relation to this symposium, the message is that whereas some environmental exposures, to pesticides, for example, may multiply the risk of developing a rare disease by 10, what should really concern the general public are absolute risks. We have been discussing influences that may prove to double the absolute risk of diseases, but they include diseases which are immensely common. This could have a considerable effect on the totality of health in a society.

Caspi: I have learnt a great deal at this meeting about why the functional context of development has to include the evolutionary and biological past; but it also has to incorporate our social *present*. Once we recognize the importance of the social present, we have to consider the fact that environments are very volatile and dynamic; yet implicitly throughout the symposium we have talked about 'the environment' as if it were a discrete entity that has persistent effects. In fact, this is not so. Consider data on poverty and income dynamics in the US. We know that children and entire family units move in and out poverty, both within and across generations. In addition, household and demographic data from the US show that by the time they reach maturity, the household composition of the children's families will have changed dramatically. Children will increasingly live with step-sibs many of whom are not even genetically related to them. There is a great deal of volatility in the environment. We need to be asking how developmental changes interact with multiple environment *changes*, not simply with a single environmental event.

Barker: This is a profound question. If in very early life you are responsive to an environment and are moulded by it, how does that leave you when the environment changes later in your life? What was advantageous formerly might become disadvantageous later.

Caspi: I don' t think we know that. We know there are a great many changes, but we don't know the consequences of those changes; so I would advocate great caution here. We have no idea how environmental *changes*, even just in terms of the nutritional context, will translate into developmental outcomes.

Richards: I agree very much with what Dr Caspi has just said, but we should also be careful about generalizing results. For example, in your paper you spoke about college education of certain groups, but in the US a much bigger proportion of the age group go to college than in the UK. Your *general* message is, I am sure, generally true, but we should be careful about generalizing specific results to other societies that organize their educational and marriage systems differently.

Wood: I am concerned with what all this means in, say, the community child health clinic. It is not possible for the health visitor or the community child health doctor to ask a mother 'how is your child's receptor today; how shall we examine your membrane?' We need to have available simple advice and clear policies—elementary, practicable nutritional policies for young children. We also need simple surveillance of growth and development at the appropriate level of access. It must be understood that the clinical world cannot be a 'reductionist' situation. We also need commonsense advice to do with the emotional hygiene of children, which can be easily, intelligently and sensitively communicated to the people (mostly parents) who actually look after children, who are people outside our medical and scientific institutions, in the community.

Thornburg: The message of the symposium is that we need a better understanding of the basic physiology of development, which could now come together by linking the molecular basis of physiological processes and medical findings together with epidemiology. If we can continue to build links as we have done here, so that we can discuss these questions across disciplines, we are likely to make progress at a faster rate than in the past. Specific funding for research projects that include intergenerational studies in addition to interdisciplinary studies is needed.

Parnas: I would agree about the need for interdisciplinary efforts. I am a member of a consortium of Danish and American researchers from a number of different disciplines. We have put up an application to create a multidisciplinary prospective cohort, because one problem of previous cohorts is that they were created by an obstetrician or paediatrician interested in an isolated problem. We are trying to create an interdisciplinary prospective cohort, with geneticists, biologists, primatologists and even dentists—to mention just some of the disciplines. Within that design, and within the limits of ethical problems, we hope to build in some prophylactic intervention measures. A manipulative component in a prospective cohort study will bring our usual inferences from correlational data closer to more causally interpretable results.

Barker: It is difficult to think that an extreme view is likely to be right in this area. One can understand John Dobbing's point that just because interesting physiological mechanisms have been described, to declare them to be important and worthy of acting on cannot be the way forward. Again, the study of outcomes alone is not necessarily going to be fertile. We know from past experience that it can be profoundly misleading unless accompanied by an understanding of biological mechanisms. We now have sufficient evidence that the early life environment is operating to affect in an important way a wide range of physiological systems that are important in later life, and is a powerful determinant of adult disease (Barker 1989). There is sufficient basis here on which to encourage exploration at both levels—disease outcomes and physiological processes.

References

Barker DJP 1989 The rise and fall of Western diseases. Nature (Lond) 338:371–372
Bitman J, Wood DL, Mehta NR, Hamosh P, Hamosh M 1986 Comparison of the cholesterol ester composition of human milk from preterm and term mothers. J Pediatr Gastroenterol Nutr 5:780–786
Brown GW, Wing JK 1962 A comparative clinical and social survey of three mental hospitals. Sociological Review, Monograph No.5, p 145– 171
Butte NF, Garza C, Smith EO, Nichols BL 1984 Human milk intake and growth in exclusively breast fed infants. J Pediatr 104:187–195
Casey CE Neville MC, Hambridge KM 1989 Studies in human lactation: secretion of zinc, copper and manganese in human milk. Am J Clin Nutr 49:773–785
Dewey KG, Lonnerdal B 1986 Infant self-regulation of breast milk intake. Acta Paediatr Scand 75:893–898
Dewey KG, Heinig LA, Mommsen LA, Lonnerdal B 1991 Maternal vs infant factors related to breast milk intake and residual milk volume: the DARLING study. Submitted
Jensen RG 1989 The lipids of human milk. CRC Press, Boca Raton, FL, p 80–89

Lucas A 1991 Programming by early nutrition in man. In: The childhood environment and adult disease. Wiley, Chichester (Ciba Found Symp 156) p 38–55

Mannan S, Picciano MF 1987 Influence of maternal selenium status on human milk selenium concentration and glutathione peroxidase activity. Am J Clin Nutr 46:95–100

Moser PB, Reynolds RD 1983 Dietary zinc intake and zinc concentrations of plasma, erythrocytes, and breast milk in antepartum and post partum lactating and nonlactating women: a longitudinal study. Am J Clin Nutr 38:101–108

National Academy of Sciences 1989 Recommended dietary allowances, 10th edn. National Academy Press, Washington, DC

National Academy of Sciences 1990 Nutrition during pregnancy. National Academy Press, Washington, DC

National Academy of Sciences 1991 Nutrition during lactation. National Academy Press, Washington, DC

Neville MC, Oliva-Rasbach J 1987 Is maternal milk production limiting for infant growth during the first year of life in breast fed infants? In: Goldman AS, Atkinson SA, Hanson LA (eds) Human lactation 3: the effects of human milk on the recipient infant. Plenum Press, New York, p 123–133

Neville MC, Keller R, Seacat J et al 1988 Studies in human lactation: milk volumes in lactating women during the onset of lactation and full lactation. Am J Clin Nutr 48:1375–1386

Paton JA, Nottebohm FN 1984 Neurons generated in the adult brain are recruited into functional circuits. Science (Wash DC) 225:1046–1048

Prentice AM, Paul AA, Prentice A, Black AE, Cole TJ, Whitehead RG 1986 Cross cultural differences in lactational performance. In: Hamosh M, Goldman AS (eds) Human lactation 2: maternal and environmental factors. Plenum Press, New York, p 13–44

Rutter M (ed) 1979/80 Changing youth in a changing society: patterns of adolescent development and disorder. Nuffield Provincial Hospitals Trust (1979), Harvard University Press, Boston, MA (1980)

Rutter M 1981 Epidemiological/longitudinal strategies and causal research in child psychiatry. J Am Acad Child Psychiatry 20:513–544

Rutter M, Pickles A 1991 Person–environment interactions: concepts, mechanisms and implications for data analysis. In: Wachs TD, Plomin R (eds) Conceptualization and measurement of organism–environment interaction. American Psychological Association, Washington, DC, in press

Smith AM, Picciano MF, Milner JA 1982 Selenium intakes and status of human milk and formula fed infants. Am J Clin Nutr 35:521–526

Tizard J 1975 Race and IQ: the limits of probability. New Behav 1:6–9

Walton A, Hammond J 1938 The maternal effects on growth and conformation in Shire horse–Shetland pony crosses. Proc R Soc Lond Ser B Biol Sci 125:311–335

Index of contributors

Subject index